GREEN CITIES

GREEN CITIES

Urban Growth and the Environment

Matthew E. Kahn

BROOKINGS INSTITUTION PRESS
Washington, D.C.

ABOUT BROOKINGS

The Brookings Institution is a private nonprofit organization devoted to research, education, and publication on important issues of domestic and foreign policy. Its principal purpose is to bring the highest quality independent research and analysis to bear on current and emerging policy problems. Interpretations or conclusions in Brookings publications should be understood to be solely those of the authors.

Copyright © 2006
THE BROOKINGS INSTITUTION
1775 Massachusetts Avenue, N.W., Washington, D.C. 20036
www.brookings.edu

All rights reserved

Library of Congress Cataloging-in-Publication data
Kahn, Matthew E., 1966-
  Green cities : urban growth and the environment / Matthew E. Kahn.
    p.   cm.
  Includes bibliographical references and index.
    ISBN-13: 978-0-8157-4816-8 (cloth : alk. paper)
    ISBN-10: 0-8157-4816-7 (cloth : alk. paper)
    ISBN-13: 978-0-8157-4815-1 (pbk. : alk. paper)
    ISBN-10: 0-8157-4815-9 (pbk. : alk. paper)
  1. Urban ecology. 2. Urbanization—Environmental aspects.
3. Cities and towns—Growth. I. Title.
HT241.K34 2006
307.76—dc22   2006021499

Digital printing

The paper used in this publication meets minimum requirements of the American National Standard for Information Sciences—Permanence of Paper for Printed Library Materials: ANSI Z39.48-1992.

Typeset in Sabon

Composition by Kulamer Publishing Services, Potomac, Maryland

Contents

Acknowledgments		vii
1	Introduction	1
2	Measuring Urban Environmental Quality	8
3	The Urban Environmental Kuznets Curve	30
4	Income Growth and the Urban Environment: The Role of the Market	50
5	Income Growth and Greener Governance	67
6	Population Growth and the Urban Environment	93
7	Spatial Growth: The Environmental Cost of Sprawl in the United States	110
8	Achieving Urban and Global Sustainability	130
References		138
Index		150

Acknowledgments

Many friends and colleagues have helped me hone this work. I made significant progress on my first draft of this book project when I visited Stanford's Economics Department during the 2003–2004 academic year. Fortuitously, several environmental economists were all at Stanford that year. Lunches with Antonio Bento, Larry Goulder, Roger Van Hafen, Jim Sanchirico, and Rob Williams helped me formulate many of the ideas presented in this book. At Tufts, I have learned a great deal about environmental issues from my colleagues Gib Metcalf and Bill Moomaw. I wrote the final draft of this book as a visiting scholar at the National Bureau of Economic Research.

This book draws on much of my research over the last fifteen years. I would like to thank my co-authors Pat Bajari, Nate Baum-Snow, Dora Costa, Michael Cragg, Denise DiPasquale, Ed Glaeser, John Matsusaka, Jordan Rappaport, Joel Schwartz, and Yutaka Yoshino. Several cohorts of skeptical students at Tufts, Columbia, and Stanford have pushed me to strengthen my arguments. I thank Brett Baden, Dora Costa, Mary Kwak, Jesse Shapiro, Phil Strahan, and Kayo Tajima for extensive comments on the entire manuscript. My editor Mary Kwak deserves special mention. I was quite lucky to work with Mary. I also owe thanks to Starr Belsky, Chris Kelaher, Terry Knopf, Celia Mokalled, and Janet Walker.

This book is dedicated to my wife Dora, our son Alex, and my parents, Martin and Carol Kahn.

CHAPTER 1

Introduction

Does growth hurt or help the urban environment? The answer, in a nutshell, is "both." Rapid growth in Asia has caused ambient particulate levels in at least twenty-five cities to rise above three times the World Health Organization's standard of 90 micrograms per cubic meter, and the mountains of refuse skirting Mexico City have become notorious worldwide. But in other parts of the world, many cities have made a dramatic quality of life comeback while continuing to grow.

In nineteenth-century New York, to take a striking example, many urbanites contended daily with fouled water, soot-darkened air, and deafening noise—to say nothing of the dead and dying horses abandoned on the streets.[1] Yet in 2004 the city's bid to host the 2012 Olympics highlighted its commitment to protecting the environment and touted New York as "a city of green."[2] During the nineteenth and early twentieth centuries, the skies above such major cities as Chicago and Pittsburgh were dark with smoke from steel smelters and other heavy industrial plants. Today, Chicago and Pittsburgh are much cleaner than they were forty years ago, and even Los Angeles has experienced a dramatic reduction in smog levels despite rapid growth in population and

---

1. Melosi (1982); Melosi (2001).
2. "Theme 5—Environment and Meteorology," *NYC2012 Bid Book*, p. 77 (www.greenorder.com/pdf/news/NYC2012_Bid_Ch5.pdf [April 2006]).

vehicle mileage. In 1880 the average urbanite in the United States had a life expectancy ten years lower than the average rural resident.[3] By 1940 this urban mortality premium had vanished.

Why do some growing cities suffer environmental degradation while others are able to preserve or even enhance their environmental quality? In recent years much work in environmental economics has focused on this question. This book draws extensively on this literature to convey what is and is not known about the environmental consequences of urban growth. While economics is called the "dismal science," economists tend to be optimistic about the consequences of economic growth.[4] Most economists have a fair amount of faith in humanity's ability to respond to incentives to economize on polluting activities. In contrast, many ecologists and environmentalists remain wary of capitalism's impact on the environment. This book does not seek to settle this dispute. Instead, its goal is to convey the excitement of an ongoing debate over the environmental consequences of market-driven growth.

Understanding the relationship between economic development and urban environmental quality is no mere academic exercise. In 2000, 80 percent of the U.S population lived in a metropolitan area, and urban growth is taking place around the world. In 1950, 30 percent of the world's population lived in cities. In 2000 this fraction grew to 47 percent, and it is predicted to rise to 60 percent by 2030.[5] Most of these cities are located in capitalist economies. With the demise of communism and China's economic transition, most urbanites live, work, and shop in free-market economies. Thus the future of urban environmental quality depends on how pollution evolves in conjunction with free-market growth.

---

3. Haines (2001).
4. However, the most optimistic recent book about environmental trends was written not by an economist but by political scientist Bjørn Lomborg, who generated headlines by arguing in *The Skeptical Environmentalist* that most environmental problems are getting better, not worse (Lomborg 2001). This punch line was backed up by 173 figures and 2,930 footnotes. Lomborg's provocative book provided detailed evidence on long-run trends, but it did not explain why some environmental indices, such as urban smog, are getting better in many cities while other sustainability indicators, such as carbon dioxide production, are getting worse.
5. United Nations, "World Population Prospects: The 2004 Revision Population Database" (esa.un.org/unpp [October 2005]).

The economists' main contribution to analysis of this issue is the concept of the environmental Kuznets curve (EKC).[6] Put succinctly, this hypothesis posits that economic development is both a *foe* and a *friend* of urban environmental quality. Economic development—especially in poorer cities—often leads environmental quality to decline, but continued development can help middle-income and richer cities solve many pollution-related problems. Why? Because as income grows, consumption and production patterns become increasingly "green" while the prospects for greener governance improve. Many studies have identified environmental indicators that fit the EKC pattern in the fifteen years since it was introduced.

But environmentalists have raised a number of important objections to the optimism implicit in the EKC. For example, some argue that even if the EKC is correct, it provides little hope to poor cities that may be trapped for a long time on the wrong side of the curve. The EKC may also have little relevance in many important areas, such as pollution problems that involve externalities on a global scale. Moreover, by focusing on changes in income, the EKC gives an incomplete picture of urban growth and its impact on environmental quality. These issues will all be taken up in this book.

What Is a Green City?

Before proceeding further, some terms should be defined. First, although I frequently highlight specific challenges facing central cities, the term *city* generally refers to a broader metropolitan area. For example, "Chicago" represents the greater metropolitan area surrounding the city of Chicago. A *metropolitan area* is a core area containing a substantial population nucleus, together with adjacent communities having a high degree of social and economic integration with that core. Metropolitan areas can comprise one or more entire counties.[7] Focusing on metropoli-

---

6. Simon Kuznets won the Nobel Prize in Economics in 1971. He studied the cross-national relationship between national per capita income and national income inequality and found evidence of a nonlinear pattern. Gene Grossman and Alan Krueger later identified a similar relationship between per capita income and pollution, as discussed in chapter 3 (Grossman and Krueger 1995).
7. See U.S. Census Bureau, "Metropolitan and Micropolitan Statistical Areas" (www.census.gov/population/www/estimates/metroarea.html [October 2005]).

tan areas makes sense because in the United States at least, a majority of people and jobs are now located within metropolitan areas but outside center cities.

Defining greenness is a tougher task. Many of us have an intuitive sense of what sets a *green* city, such as Portland, Oregon, apart from *brown* urban centers, like Mexico City. Green cities have clean air and water and pleasant streets and parks. Green cities are resilient in the face of natural disasters, and the risk of major infectious disease outbreaks in such cities is low. Green cities also encourage green behavior, such as the use of public transit, and their ecological impact is relatively small.

Can this subjective definition of a green city be translated into objective indicators of urban environmental quality? Chapter 2 examines efforts in three different fields to do just that. Ecologists emphasize the importance of tracking the size of a city's ecological footprint. This approach focuses on how much people consume and how much carbon dioxide is produced as a byproduct of urban consumption and production. Public health experts focus on the health consequences of exposure to local air pollutants, dirty water, and other environmental factors that promote disease. Based on this approach, a city is considered green if the incidence of environmentally linked diseases is relatively low. Finally, many economists evaluate the urban environment by examining differences in real estate prices across cities at a point in time or for the same city over time. If home prices are much higher in San Francisco than in Detroit, this suggests that people prefer to live in San Francisco—in part because of its superior environmental quality. Otherwise, mobile households could enjoy a "free lunch"—a cheap house with no sacrifice of quality of life—by moving from San Francisco to Detroit.

Each approach has advantages and disadvantages. Equally important, the three approaches can lead to different conclusions about urban environmental quality. For example, some cities boast low local pollution levels and a high quality of life but generate relatively high levels of greenhouses gases. Are these green cities? The answer to this question depends on how one prioritizes local urban challenges, such as smog, versus longer-run global challenges, such as climate change. Chapter 2 addresses this problem by suggesting how various indicators can be combined to create a "green city" index. Although we currently lack the data necessary to construct such an index, this exercise helps clarify what we mean when we say that a city is green. My own view is that a green city should score high marks when graded on both a local and a global scale. In other words, in addition to enjoying the benefits of clean air and water, its residents should avoid imposing negative externalities on people who live beyond the city's borders.

The Two Faces of Growth

How does growth affect a city's prospects for becoming more—or less—green? Chapter 3 takes a first cut at this problem by providing an overview of the environmental Kuznets curve, including a discussion of its history and some examples of environmental indicators that follow the EKC pattern—that is, first deteriorating and then improving as per capita income grows. This chapter briefly describes the main channels through which income growth affects environmental quality, as well as several key factors that can alter the shape of the EKC. In addition, it presents several limitations to the hypothesis, including concerns raised by environmentalists.

Income Growth and the Urban Environment

Chapters 4 and 5 explore the mechanisms behind the EKC in greater detail. Chapter 4 examines how income growth can enhance urban sustainability—even in the absence of government intervention—by promoting changes in urban consumption and production patterns. Richer urbanites, for example, are more likely to purchase green products and amenities, such as newer vehicles that pollute less per mile. In addition, as wages and education rise, a city's industrial composition often changes. Heavy manufacturing tends to be priced out of richer cities, giving way to relatively low-pollution industries, such as services and finance. These sectors rely on access to a well-educated workforce, which gives them a financial incentive to participate in efforts to preserve a city's quality of life.

Chapter 5 moves beyond the market to investigate how income growth affects the prospects for greener urban governance. Economic development can potentially increase both the demand for and supply of environmental regulation. As residents become wealthier, they have an increased desire to live in a high quality of life area. As a result politicians have stronger incentives to invest in green policies. They also have greater access to policy resources as a city's income grows. Chapter 5 addresses these issues by examining recent efforts to confront major urban environmental challenges in the United States. It also highlights regulation's intended and unintended effects.

Population Growth and the Urban Environment

By focusing on growth in income, the environmental Kuznets curve hypothesis neglects other key aspects of urban growth. Chapters 6 and 7 remedy this oversight by exploring the relationships among population growth, population density, and spatial growth in cities in the developing world and the United States.

Chapter 6 focuses on the relationship between urban population growth and environmental quality. In many developing nations, cities act as magnets, drawing people out of the countryside to urban jobs. Inevitably, a growing urban population consumes more resources and generates increased waste. In the absence of effective policies to counteract these effects, fast-growing cities in developing countries experience sharp increases in all types of pollution. Ongoing research attempts to measure the quantitative size of these effects.

Population growth can also contribute to urban environmental problems in other ways. Growth often increases urban income inequality and ethnic heterogeneity. In a highly diverse city, different interest groups may disagree over what is "good public policy" and who should pay for these policies. Chapter 6 investigates some of the effects this dynamic can have on the urban environment.

Spatial Growth and the Urban Environment

While many cities in developing countries suffer environmental problems due to high population density, in the United States, the fastest growth is taking place in low-density, car-friendly metropolitan areas. According to U.S. census data, in 2000, across all metropolitan areas in the United States, 53 percent of employed heads of households lived in detached homes and commuted to work in private vehicles. Environmentalists argue that this suburban sprawl is socially costly. They claim that the pursuit of the "American Dream"—often defined as owning two cars and a large suburban house—translates, in aggregate, into an enormous ecological footprint. Chapter 7 presents new evidence on how suburbanization affects household resource consumption and urban sustainability.

Beyond City Limits

While continuing to grapple with local environmental problems, many cities also expect to face new challenges as a result of climate change. For example, coastal cities, especially those closer to the equator, will face a greater risk of flooding and extreme heat. Does Hurricane Katrina's blow to New Orleans foreshadow future urban impacts? If climate change increases the frequency and severity of natural disasters, the answer may be yes.

In theory cities could help head off these problems. After all, cities are leading centers of idea generation. Urban centers may incubate new technologies that could weaken the link between economic activity and greenhouse gas production. But cities also play a major role in increasing the risk of climate change by generating greenhouse gases, such as carbon dioxide. Since reducing emissions is costly, and the benefits of doing so are shared with the rest of the world, each city has few incentives to limit greenhouse gas production on its own. This is a classic example of the free-rider problem.

Will urban growth simply exacerbate the problem of climate change, or can it help address this challenge? In the short term, it seems likely to make the problem worse. Urban growth fosters economic development by encouraging trade and specialization. As incomes rise, households consume more energy at home, at work, and on the road. However, urban growth can also have potentially offsetting effects. For example, urbanization can reduce population growth at the national level and facilitate emission-reducing technological advance. Does this suggest that greenhouse gas production is likely to follow the pattern of the EKC? Chapter 8 reviews the evidence on this question and concludes by asking what climate change is likely to mean for cities around the world.

CHAPTER 2

Measuring Urban Environmental Quality

When asked to name a green city, many people would say San Francisco or Vancouver, but few would say Houston. Why? What determines whether a city should be considered green or brown? What yardsticks should be used when comparing cities or creating city rankings?

Ecologists, public health experts, and economists approach this task in different ways. Ecologists focus on measuring changes in natural capital stocks over time. Public health researchers seek to measure the excess morbidity and mortality risk associated with diseases caused by pollution exposure.[1] Economists examine urban home prices and wages to see whether people are paying a premium—measured in higher home prices and lower wages—to live in a specific city. Each of these methods provides useful clues concerning the environmental costs of growth. This chapter investigates the strengths and weaknesses of each approach.

Ecological Footprints

Ecologists have captured the public's attention with a measuring stick called the ecological footprint. This approach measures the resources

---

1. Morbidity refers to the incidence of a disease within a population; mortality refers to the death rate associated with that disease.

consumed and the waste produced by a given entity and translates this figure into the land and water area required to support this level of activity. An ecological footprint can be constructed for an individual or for population groupings, such as cities, nations, or the entire planet. A major strength of this approach is that it provides an intuitive formula for converting day-to-day individual choices into an aggregate measure of demand for natural capital.

As an example, consider the online footprint calculator sponsored by two environmental nonprofits, Redefining Progress and Earth Day Network.[2] This tool estimates an individual's ecological footprint based on answers to the following thirteen questions:

—How often do you eat animal-based products?

—How would you describe your average daily caloric food intake?

—How much of your purchased food is thrown out rather than eaten?

—A significant portion of the energy cost of food production is spent on transporting food from harvest to market and for processing, packaging, and storage. Purchasing locally grown, in-season, unprocessed food can greatly reduce the need to expend energy in food production. How much of the food that you buy is locally grown, unprocessed, and in season?

—How much do you drive each year, on average (either as a driver or passenger)?

—On average, how often do you drive with someone else (either in your car or theirs)?

—How many miles per gallon does your car get?

—On average, how many miles do you travel on public transportation?

—How many hours each year do you spend flying?

—How many people live in your home?

—How big is your home?

—Does your home purchase electricity from a "green" electricity provider (for example, solar, wind, microhydroelectric)?

—Do you use energy-efficient appliances and light bulbs?

After completing this questionnaire, I learned that my ecofootprint measures 91.4 percent of an average American's footprint and that if the rest of the world were to enjoy my standard of living, it would require

---

2. Redefining Progress and Earth Day Network, "Ecological Footprint Quiz" (www.ecofoot.org [October 2005]).

5.2 Earths to support the present human population. In other words, if the residents of developing countries such as China and India choose consumption patterns similar to the current "American Dream," then in a matter of decades (assuming per capita gross national product in these countries continues to grow at current rates), the demand for natural capital will vastly exceed the supply. This prognosis is unlikely to change even if world population growth levels off. As Jared Diamond writes in his latest bestseller, *Collapse,* "The larger danger that we face is not just of a two-fold increase in population, but of a much larger increase in human impact if the Third World's population succeeds in attaining a First World living standard."[3]

Similar conclusions emerged from a recent study that examined long-run trends in the world's ecological footprint. A team of researchers led by Mathis Wackernagel measured global consumption along six dimensions: growing crops for food, animal feed, fiber, oil, and rubber; grazing animals for meat, hides, wool, and milk; harvesting timber for wood, fiber, and fuel; marine and freshwater fishing; developing and maintaining infrastructure for housing, transportation, industrial production, and hydroelectric power; and burning fossil fuel. They then calculated the land area required both to carry out these activities and to absorb the resulting wastes. Based on this approach, they concluded that "humanity's load corresponded to 70 percent of the capacity of the global biosphere in 1961 and 120 percent in 1999."[4] Increased consumption of fossil fuels accounted largely for this growth.[5] In 1961 energy consumption represented roughly 25 percent of the world's ecological footprint, but by 1999 that figure had risen to 50 percent.[6]

Footprints and Urban Growth

Calculating a city's ecological footprint can provide a good sense of recent trends in resource consumption. But what conclusions can be reached regarding the relationship between urban growth and environmental quality? If a city's population doubles from 1 million to 2 million,

---

3. Diamond (2005, p. 511).
4. Wackernagel and others (2002).
5. Burning fossil fuels generates large amounts of carbon dioxide. Absorbing this waste requires huge investments in carbon sequestration—primarily through the planting of forests—which is very land intensive.
6. Wackernagel and others (2002).

will its footprint double in size? Will this development cause the country's footprint to increase? The answer to both questions is probably no.

Consider, for example, urban growth arising from rural to urban migration. In this case there are two offsetting forces to consider. Urbanites earn and consume more than rural households, which suggests that urbanization might increase a nation's ecological footprint. However, urban households have smaller families than rural households. If overall national population growth slows because of urbanization, then the national footprint could actually shrink.

In addition, well-functioning markets can help moderate the environmental impact of urban growth. As cities grow, the laws of supply and demand will cause the prices of scarce market goods, such as land, water, and other forms of natural capital, to rise. Rising prices will serve as incentives to economize on resource consumption and to develop and adopt green technologies.[7] These activities will limit the growth of the city's footprint, not due to any particular concerns about sustainability but due to the narrow pursuit of self-interest.

Even the *expectation* of future price increases can signal producers and consumers to alter their behavior. For example, assume that the owners of oil reserves believe that the price of oil will rise faster than the interest rate, perhaps due to rapid growth in the number of people who can afford cars. Under these conditions they have an incentive to store oil now and sell it in the future at a higher price. In the short term, this reduction in supply will raise prices and encourage consumers to use public transit more often and economize on vehicle trips. In the longer run, vehicle manufacturers may respond to higher fuel prices by building more fuel-efficient cars.[8] Why would profit-maximizing firms adopt such an approach? Around the world there are many car makers. If one

---

7. If resource prices do not adjust to equate supply and demand, then rising demand can lead to resource shortages. If a shortage did take place, then this would increase the demand for a political response involving conservation and higher resource prices.

8. An oil analyst named Matt Simmons has suggested that Saudi Arabia may actually be trying to head off such behavior. Based on his research on Saudi oil reserves, he has accused Saudi Arabia of overstating its true reserves. Why would Saudi Arabia lie and tell the world that it has plenty of oil? Such a declaration would reduce consumer demand for greener cars and weaken automakers' incentives to invest in fuel-efficient vehicles. Peter Maass, "The Breaking Point," *New York Times*, August 21, 2005, p. 30. See also Simmons and Company International, "Recent Speeches and Papers Presented by Matthew R. Simmons" (www.simmonsco-intl.com/research.aspx?Type=msspeeches [May 2006]).

of these companies bets that real oil prices will rise and invests in green technologies to prepare for that day, then this company could seize the market when gas prices rise to $10 a gallon.

Government intervention can hasten this adaptive process. The creation of a "carbon tax" would strengthen consumers' incentives to economize on the production of greenhouse gases—whether by buying smaller cars, driving less, or bundling up in winter and turning the thermostat down. The potential impact of this measure is suggested by consumer responses to previous increases in the price of gasoline. For example, in the wake of Hurricane Katrina, as gas prices in the United States rose over $3 a gallon, sales of fuel-efficient cars like the Honda Civic increased and demand for SUVs dropped.[9] This suggests that even growing cities and nations could shrink their ecological footprints by adopting incentive systems that encourage economizing on natural capital.

The Ehrlich-Simon Bet

Many economists believe that due to such behavioral responses, the world is unlikely to run out of key nonrenewable resources, such as oil.[10] This is by no means a universal view. Recently, for example, there has been a great deal of interest in the concept of Hubbert's peak. In 1956 M. King Hubbert, a prominent geophysicist, predicted correctly that U.S. oil production would peak in the early 1970s. Using Hubbert's methods, other analysts have predicted that world oil production will

---

9. Sholnn Freeman, "Truck and SUV Sales Plunge as Gas Prices Rise," *Washington Post*, October 4, 2005, p. D01.

10. Ironically, the world may be more likely to run out of renewable resources, such as fish in the oceans, due to a problem known as the tragedy of the commons. As Robert Stavins explains, "It is the renewable resources that in some cases are very much exhaustible, not because of their finiteness, but because of the way they are managed. The problem typically is not physical limits on resource availability; on the contrary, improper incentives and inadequate information are more often the source of the declining stocks. Thus, the reason why some resources—water, forests, fisheries, and some species of wildlife—are threatened, while others—principally minerals and fossil fuels—are not is that the scarcity of nonrenewable resources is well reflected in market prices" (Stavins 1992). However, not all researchers agree on this point. For example, one study using data from India has found that although population and income growth result in increased demand for lumber, the supply of forests actually increases and deforestation slows (Foster and Rosenzweig 2003). Anticipated demand for these renewable resources provides incentives for those who control access to forests to manage this valuable resource wisely.

peak within a decade.[11] Traditional media outlets, such as the *New York Times,* and blog sites, such as the Oil Drum, have devoted ample attention to this claim, which is generally cited as evidence that the world faces an impending energy crisis.[12]

However, the outcome of the famous bet between Paul Ehrlich and Julian Simon suggests that this need not be the case. Ehrlich, a Stanford University ecologist, saw a finite supply of natural resources and rising world demand due to increased population and income. Therefore he predicted that the prices of these resources would increase. Simon, an economist at the University of Maryland, countered that technological advance would bring about increases in supply of these resources and would increase substitution possibilities such that demand could actually decline. As a result, he argued, real commodity prices could fall over time. In 1980 the two men decided to back their words with action:

> Simon then offered Ehrlich a bet. Ehrlich could choose any five raw materials he wanted. Simon sold Ehrlich an option to buy an amount of each raw material worth $200 in 1980 dollars. If the prices increased over the next ten years, Simon would pay Ehrlich; however, if the prices decreased over the same time period, Ehrlich would have to pay Simon. Ehrlich chose five metals: copper, chrome, nickel, tin and tungsten. The bet was on. Ten years later, after adjusting for inflation, just as predicted the prices of all five metals went down. Ehrlich had lost.[13]

Ehrlich lost the wager because he assumed that current consumption and production trends would persist. Worst-case scenarios that rely on this assumption can be highly misleading, given the potential for technological, regulatory, and behavioral change.

Healthy Cities—and Sick Ones

A second approach to measuring urban environmental quality has been developed by researchers in the field of public health. This approach

---

11. Deffeyes (2001).
12. See www.theoildrum.com.
13. Daniel B. Botkin and Edward A. Keller, "Environmental Debates: Part III: Paul Ehrlich vs. Julian Simon—Have We Reached the Limit?" In *Environmental Science: Earth as a Living Planet,* 5th ed., Student Companion Site (bcs.wiley.com/he-bcs/Books?action=index&itemId=047148816X&bcsId=2054 [March 2006]).

judges a city to be "brown" if environmentally related health problems are above average or rising over time. Suppose that Pittsburgh residents suffer more health problems in 2003 than in 1993 and that they are also sicker than residents of roughly similar cities, such as Cleveland. Suppose further that Pittsburgh residents are the same age and ethnicity as residents of these other cities and that their diet, exercise, and smoking habits are all roughly comparable. In this case a public health expert would conclude that Pittsburgh's environment is contributing to its population's health problems—especially if Pittsburgh's citizens do not suffer from higher rates of diseases with no known environmental component.

Measuring Pollution's Effects

A major research focus in public health has been to calculate credible estimates of the improvements in urban health that can be achieved by reducing pollution—or alternatively, of the harm that is done by rising pollution levels. This is not always as straightforward as it may seem. Health problems could be high in cities with high pollution levels for reasons unrelated to pollution exposure. For example, poorer people who are more likely to smoke might be more likely to live in highly polluted cities. If this is the case, then a researcher who naively compares the incidence of health problems in low-pollution versus high-pollution cities will overestimate the impact of pollution.

To measure such effects accurately, researchers typically seek a "natural experiment" in which there is a sudden and unexpected change in pollution levels. For example, Michael Ransom and C. Arden Pope exploited the fact that labor strikes lead to intermittent operation of a steel mill in a mountain valley in central Utah.[14] This provides a unique opportunity to measure the external health costs of air pollution, with a nearby valley serving as a control. After analyzing data on hospital admissions and daily deaths for the two valleys, the researchers found that hospital admissions for respiratory diseases increase significantly when the mill is in operation and air pollution is high. Based on the steel mill's contribution of sulfates to the environment, their analysis concludes that the mill's operation accounts for twenty to seventy-five deaths a year.

The Olympic Games offer another natural experiment for testing how reductions in local air pollution affect urban public health. In 1996 the Olympics took place in Atlanta. A team of researchers took advan-

14. Ransom and Pope (1995).

tage of this event to compare the seventeen days of the Olympic Games (July 19 through August 4) to a baseline period consisting of the four weeks before and after the event.[15] During the games peak traffic density measurements were 22 percent lower than in the comparison periods. As a result the peak one-hour level of ambient ozone fell to 50–100 parts per billion from a predicted value of about 70–120 parts per billion in the comparison period, and ambient particulate matter levels were much lower than usual.[16] These reductions in pollution translated into significant public health gains. During the Olympic Games, in the five central counties of metropolitan Atlanta, the number of asthma-related hospitalizations, emergency department visits, and urgent care center visits for children aged one to sixteen decreased by 41.6 percent.

Recessions provide yet another opportunity to study the impact of exogenous changes in air pollution. The 1981 recession in the U.S. Rust Belt led to a large decline in dirty manufacturing jobs in some cities and left employment—and consequently air pollution—in nonmanufacturing cities largely unchanged. Kenneth Chay and Michael Greenstone used this variation to identify the effect of air pollution on infant mortality.[17] Their analysis concluded that a 10 microgram per cubic meter reduction in particulates reduces infant mortality by 55 infant deaths per 100,000 live births at the county level.[18] Given that some U.S. cities, such as Los Angeles, Chicago, and Pittsburgh, have experienced a reduction of over 30 micrograms per cubic meter in this environmental indicator over the last thirty years, this suggests that very large public health gains have taken place.

Measuring Pollution's Cost

A leading advantage of the public health approach is that it provides a framework for measuring the costs imposed by environmental problems. Suppose that public health research concludes that urban pollu-

15. Friedman and others (2001).
16. Ozone and particulates are two of the six ambient air pollution criteria regulated under the Clean Air Act. The other measures of air pollution are carbon monoxide, sulfur dioxide, lead, and oxides of nitrogen.
17. Chay and Greenstone (2003).
18. "'Particulate matter,' also known as particle pollution or PM, is a complex mixture of extremely small particles and liquid droplets. Particle pollution is made up of a number of components, including acids (such as nitrates and sulfates), organic chemicals, metals, and soil or dust particles." See U.S. Environmental Protection Agency, "Particulate Matter" (www.epa.gov/oar/particlepollution [March 2006]).

tion causes the average urbanite to experience one extra sick day a year and increases his or her risk of dying that year by 0.005 percentage points. In a city of 1 million people, this would mean that pollution exposure results in the loss of 1 million work days and fifty statistical lives per year.[19] An economist would take these estimates and translate them into a dollar cost of urban air pollution.

Calculating the cost of lost days of work is fairly straightforward. If an individual is sick for a day, that person loses a day of earnings. So, if the average worker earns $100 a day, the loss of 1 million work days would amount to $100 million a year. In contrast, placing a value on the extra mortality risk associated with air pollution is more complicated. Economists typically use real-life tradeoffs to put a dollar value on a statistical life. For example, by collecting wage data on comparable low-risk and high-risk jobs (bartenders versus construction workers, for example), they can identify the wage premium that individuals receive for working under conditions that involve a greater risk of death.[20]

One criticism of using wage data to infer the willingness to pay to avoid a small chance of death is that those who actually like to gamble may choose to work in risky jobs. If risk-loving workers took all the risky jobs, then one would never observe the wage premium that would be required to lure a risk-averse worker to take a gamble. In the more realistic case where workers differ with respect to their taste for risk, this approach is likely to underestimate the value of a statistical life because risk-loving workers require a smaller wage premium for doing dangerous jobs.

Recognizing this problem, economists have shown creativity in devising alternative ways to infer how much people value marginal increases in safety. For example, Orley Ashenfelter and Michael Greenstone have used state-mandated changes in highway speed limit laws to infer the value of life.[21] (When a state raises its speed limit, it implicitly trades off hours commuting, which decline, against traffic fatalities, which increase.) Through such methods, economists have generally reached the conclusion that the value of a statistical life in the United States can be estimated at between $2 million and $6 million.

19. Economists use the term "statistical life" because it is impossible to identify ex ante who will be the unlucky fifty people in the million-person city who will die from the pollution exposure.
20. See Viscusi (1993).
21. Ashenfelter and Greenstone (2004).

This approach is controversial. Many people are uneasy about placing a "value" on life. Moreover, the wage methodology has been criticized on equity grounds. In developing countries where wages are lower, smaller risk premiums are associated with dangerous jobs, which leads economists to reach lower estimates of the value of a statistical life. One World Bank study using wage data from India in the late 1990s estimated a value of life at only $300,000.[22]

But in a world of finite resources, such calculations can help policymakers make better decisions about where to spend city funds. While some environmental problems, such as high ambient ozone levels, cause illness and lost days of work, others can result in an increase in the probability of death. High particulate levels and contaminated water, which promotes the spread of diseases like cholera, fall into this second group. Estimating the cost of these risks can facilitate urban health planning, as well as comparisons among different cities.

Protective Strategies

Efforts to use such comparisons to rank cities in terms of greenness are complicated by the fact that many factors help determine how pollution affects a city's health. Geographic factors, for example, can sharply increase or reduce a city's vulnerability to dangerous emissions. Polluting factories typically do less damage in cities that benefit from the cleansing effects of frequent rain and wind than in arid, mountain-locked cities like Santiago or Mexico City. Recent World Bank research has documented this effect. Using total ambient suspended particulate data for 118 monitoring stations around the world's cities, Susmita Dasgupta and coworkers explored the relative importance of urban population, governance, national income, and geographic factors in determining local air pollution levels. Based on this analysis, they concluded:

> Among the four determinants, locational advantage (i.e., climate and wind patterns) clearly has the greatest impact across its range in the sample data. Holding income, governance, and population density constant at "worst-case" levels, changing locational advantage from low to high reduces predicted air pollution from 437 to 131 µg/m^3. This result suggests that geographic factors alone are sufficient to determine whether a poor, overcrowded, poorly-

22. Alberini and others (1999).

governed city will suffer from crisis-level air pollution, or experience pollution near the upper bound for air pollution in OECD cities.[23]

Cities blessed with growth-friendly geography suffer much less public health damage for a given level of economic activity. No public policy can alter this important exogenous effect.

But city officials can also implement relatively low-cost policies to reduce the impact of pollution. Many cities, ranging from Los Angeles to Mexico City, alert residents through radio and Internet announcements when smog levels are expected to be dangerously high.[24] If people respond to this information by staying inside and avoiding strenuous outdoor activity, then smog-related health problems will be lower than might otherwise be the case. Such "information regulation" breaks the link between outdoor pollution and population exposure and raises the possibility that the public health approach might rate a city as "green" even when there are severe ambient environmental problems.

Economists have studied the impact that initiatives such as the Los Angeles "smog day" alert have on public health. A smog alert takes place when forecasted ozone levels exceed 0.20 parts per million. A recent study compared hospitalization rates for respiratory conditions on smog alert and non–smog alert days and found that fewer people were hospitalized for asthma on smoggier (that is, smog alert) days.[25] While the study cannot prove that smog alert announcements *caused* Angelenos to change their behavior, it suggests strongly that collecting and distributing information concerning local environmental quality can have a significant positive impact on public health. However, one drawback to this strategy is that public health announcements mainly benefit wealthier and more educated residents. Researchers who have examined

---

23. Dasgupta and others (2004).
24. As defined by the *Columbia Encyclopedia*, smog is "dense, visible air pollution" of two types: "The gray smog of older industrial cities like London and New York derives from the massive combustion of coal and fuel oil in or near the city, releasing tons of ashes, soot, and sulfur compounds into the air. The brown smog characteristic of Los Angeles and Denver in the late 20th cent[ury] is caused by automobiles. Nitric oxide from automobile exhaust combines with oxygen in the air to form the brown gas nitrogen dioxide. Also, when hydrocarbons and nitrous oxides from auto emissions are exposed to sunlight, a photochemical reaction takes place that results in the formation of ozone and other irritating compounds." See *Columbia Encyclopedia*, 6th ed. (www.bartleby.com/65/sm/smog.html [March 2006]).
25. Neidell (2004).

time diaries on low-smog and high-smog days in California have found that the most educated are most likely to spend more time indoors on high-pollution days.[26]

The urban poor are also less likely to have access to a broad range of strategies that limit their vulnerability to pollution. While they may have equal opportunities to exercise or forgo smoking, the poor can rarely afford quality health care or a wide variety of protective products, ranging from vitamins to vacation homes where they can take refuge during peak pollution months. This discrepancy raises an interesting issue concerning social justice and the public health approach for measuring urban environmental quality. Imagine a city where 10 percent of the population is poor and the remaining 90 percent are middle class or richer and thus are able to afford self-protection strategies. Imagine further that middle-class and rich residents suffer zero sickness per year while the average poor person in the city is sick for sixty days a year due to pollution exposure. In this case the average resident is sick for six days a year. A researcher with only city-level average data might conclude that the impact of pollution is low whereas a sociologist who surveys a low-income neighborhood in this city would reach a very different conclusion.

In a city populated with poor and rich people, should judgments of urban health be based on a fictitious "average" person? Or should they follow the standards of John Rawls in requiring that a fair society focus on raising the well-being of the worst off?[27] If the second approach is chosen, then expensive self-protection strategies have little impact because they do nothing to improve the lot of those at the bottom of the income distribution.

Assessing the Public Health Approach

The uneven availability of self-protection strategies highlights one of the main weaknesses of the public health approach: namely, the ability to offset pollution exposure can lead to counterintuitive conclusions regarding the quality of the urban environment. Suppose that an industrial city's economy is booming. This raises local income levels and pollution levels. If richer people quit smoking and eat better diets, then

---

26. Bresnahan, Dickie, and Gerking (1997).
27. Rawls (1999).

local health levels could actually improve despite the increased soot and smog.[28] People who can withstand pollution may also be more likely to move to polluted places. In an extreme case, if a group of "supermen" and "superwomen" with terrific lungs lived in the most polluted cities because they could cope with the smoke, and they enjoyed the cheap housing affordable in such an unpleasant city, a naive researcher would conclude that exposure to pollution has no impact on public health.

One way around this problem might be to use investments in self-protection as an alternative measure of urban "greenness." Expenditures on vitamins or doctor visits represent market purchases that offset pollution exposure. If two similar people who live in two different cities are making very different payments for self-protection, then one might conclude that the person who pays more lives in the less sustainable city. However, in practice, it is often difficult to determine what constitutes "defensive expenditure." Does a second home represent such an investment—or simply a luxury?

Expensive Cities—and Cheap Ones

A third approach to measuring environmental quality uses differences in housing prices across cities to measure differences in urban quality of life. People are used to the fact that a Mercedes costs more than a Toyota due to differences in perceived quality. Economists apply the same intuition to urban real estate markets. High real estate prices indicate that an area is a desirable place to live, in part because of its environmental quality.

Suppose that a city has earned a reputation as a "brown city" where industrial polluters are degrading local quality of life. If both current and potential residents are offended by how the city looks and smells and are aware of the resulting risks to public health, then households are likely to "vote with their feet." Current residents will move out of the city, and potential entrants will choose not to move in. As a result

28. Simply eating a healthier diet rich in proteins increases the body's ability to combat any environmentally induced disease (Lee 1997). Partially as a result, historical evidence indicates that public health threats in cities have plummeted as income increased. As incomes rise, the combination of better diets, lower density, more information about the benefits of washing and hygiene, and better public health infrastructure generally leads to a sharp decline in the urban disease environment. For example, death from tuberculosis in San Francisco decreased between 1900 and 1927 from 325 per 100,000 to 100 per 100,000 (Craddock 2000).

housing prices will fall. At the same time, cities offering a higher quality of life will experience in-migration, which will bid up land prices (and bid down local wages).[29] Prices will continue to adjust until migrants are just indifferent between living in the nicer city and the less pleasant one. In order for this condition to hold, the city with lower quality of life must offer lower rents and higher wages than the more attractive one. Economists call this implicit payment for local public goods a compensating differential. Households face a trade-off: if they want to live in a nicer city, they must pay more for housing. There is no free lunch.

Some cities, such as San Francisco, consistently feature higher real estate prices than others, such as Detroit or Houston. These price differences largely reflect differences in quality of life. To document the large and persistent differences in home prices for observationally similar housing structures, I have used data from the 2000 U.S. Census of Population and Housing, which collects information on self-reported home values and housing structure characteristics. Table 2-1 shows median home prices for six-room homes in 1980 and 2000 for twenty-seven major metropolitan areas. The differences in home prices across the country are quite large. In 2000, for example, the median price of the same housing structure was $75,000 in Pittsburgh and $225,000 in Los Angeles.

What Do People Pay For?

While it is easy to establish cross-city differences in home prices and wages, a more ambitious research agenda seeks to decompose these price differences into objective amenity differences and identify the most important components of urban quality of life. For example, based on home price differentials around the world, several researchers have documented the importance of climate as an urban amenity. Recent studies

---

29. While this point would appear to be quite intuitive, consider this quote from Jared Diamond's *Collapse*: "Thus, environmental and population problems have been undermining the economy and quality of life in Southern California. They are in large measure ultimately responsible for our water shortages, power shortages, garbage accumulation, school crowding, housing shortages and price rises and traffic congestion" (2005, p. 503). If Los Angeles is truly such a mess, why is its population growing and home prices increasing? Diamond may be correct that this city could be even nicer if such congestion issues could be resolved (I will return to these issues in chapter 5), but the typical household compares Los Angeles as it is today to other alternatives, such as Houston or Detroit.

Table 2-1. *Home Prices across Major Metropolitan Areas, 1980 and 2000*
Median prices in thousands of 1999 dollars

Metropolitan statistical area	Value in 1980	Value in 2000
Atlanta	88.8	112.5
Baltimore	88.8	95.0
Boston	109.7	225.0
Chicago	130.6	137.5
Cleveland	109.7	112.5
Dallas	109.7	85.0
Denver	141.1	162.5
Detroit	88.8	112.5
Ft. Lauderdale	130.6	112.5
Houston	109.7	75.0
Kansas City	88.8	95.0
Los Angeles	198.6	225.0
Miami	120.2	137.5
Milwaukee	130.6	137.5
Minneapolis	130.6	137.5
New York City	109.7	187.5
Philadelphia	67.9	95.0
Phoenix	120.2	112.5
Pittsburgh	88.8	75.0
Portland	120.2	162.5
Riverside, Calif.	141.1	137.5
St. Louis	88.8	95.0
San Diego	177.7	225.0
San Francisco	198.6	275.0
Seattle	141.1	187.5
Tampa	88.8	95.0
Washington, D.C.	141.1	137.5

Source: Author's calculations based on the median self-reported home value for single detached homes with six rooms, using the Integrated Public Use Microdata Series database, which reports home values in categories. See Minnesota Population Center, University of Minnesota, "Integrated Public Use Microdata Series" (www.ipums.umn.edu/usa/ [May 2006]).

have documented the importance of temperate climate as a key determinant of cross-city real estate prices in both Italy and Russia.[30] Perhaps not surprisingly, rents are lower in frigid Siberia than in Moscow, where it is merely extremely cold. Similarly, using census data from the United

30. Maddison and Bigano (2003); Berger, Blomquist, and Sabirianova (2003). The recent European Union expansion and labor market integration offer another interesting test of hedonic capitalization. As European Union integration continues, will home prices rise in European cities with more temperate climates relative to colder northern cities? Or, due to cultural and language barriers, will arbitrage opportunities persist such that lucky migrants can enjoy improved amenities without seeing their wages fall and rents rise?

States, Dora Costa and I show how the implicit price of climate has changed from 1970 to 1990.[31] Housing prices are higher in cities with higher January temperatures, lower July temperatures, and lower rainfall throughout this entire period, but the differential has increased over time. In 1970 a person had to pay a premium of $1,288 a year (in 1990 dollars) to purchase San Francisco's climate rather than Chicago's. In 1990 this premium had risen nearly sixfold to $7,547 a year.

Within a major city, communities differ with respect to their quality of life. Some communities have low crime levels while others have high pollution levels. Denise DiPasquale and I use 1990 census microdata for Los Angeles to measure how much households pay for local public goods.[32] In other words, home prices are regressed on neighborhood attributes, such as crime and pollution levels. This technique allows us to examine how much of a home's price is due to its structure versus the attributes of its local community. All else equal, home prices are 3 percent lower in communities where the Clean Air Act standard for ozone smog is exceeded an additional ten days a year.

Smog is not the only environmental "public bad" that people will pay to avoid. Using 1980 and 1990 data for all U.S counties, Kenneth Chay and Michael Greenstone found that a 10 percent reduction in total suspended particulates increased home prices by 3 percent.[33] Similarly, when a hazardous waste site becomes eligible for a Superfund cleanup, median home prices nearby increase by 6 percent.[34] In contrast, when a hazardous site is first marked as requiring a Superfund cleanup and placed on the National Priority List, homes close to the site decline in value. The marginal price of an extra mile of distance from the site peaks at $2,364 and declines to zero at a distance of 6.2 miles.[35] Homes near industrial sites, commercial development, and sewage treatment facilities all sell at a discount whereas easy access to green space pushes house values up.[36]

31. Costa and Kahn (2003b).
32. DiPasquale and Kahn (1999).
33. Chay and Greenstone (2005). Cities and communities that offer high environmental quality are likely to have higher real estate prices for two reasons. First, the environmental quality will be positively capitalized into home prices. Second, high environmental quality areas tend to attract more educated residents. Urban researchers have documented that home prices appreciate in areas where the educated live (Rauch 1993; Bajari and Kahn 2005).
34. Greenstone and Gallagher (2005).
35. Kohlhase (1991).
36. Lee and Linneman (1998); Necheyba and Walsh (2005).

Assessing the Compensating Differentials Approach

Real estate prices provide a direct report card on the evolution of urban quality of life. The strength of this approach to ranking cities is that it is based on what economists call "revealed preference." From collectable market data, it is possible to learn about the actual trade-offs that urbanites are willing to make between market consumption and nonmarket local public goods, such as climate and pollution.

But this approach also has limitations. For one, real estate prices reflect not only demand but also supply. Recent urban economic research has documented large differences in supply conditions across metropolitan areas. Edward Glaeser and two of his colleagues have shown that regulation in major cities in California and the Northeast, including Boston and New York City, has a significant impact on reducing the supply of housing.[37] For example, in New York City, opponents of new buildings sometimes appeal to local zoning boards to block construction, arguing that it would lower the value of their homes by cutting off sunshine and blocking their views of Central Park. The net result of such lawsuits is to reduce the supply of new housing and drive up prices relative to less regulated markets. In cities where it is difficult to build new housing, increases in demand translate into higher market prices while in cities such as Houston, where it is easy to build, increased demand has relatively little impact on local home prices.

Equally important, the compensating differential approach is only useful to the extent that residents both know and care about environmental threats and have the ability to act on that knowledge. This method cannot be used to value environmental threats that people may not be aware of. For example, suppose that living close to power lines

---

37. Glaeser, Gyourko, and Saks (2005) argue that several factors have contributed to the man-made scarcity of housing. Judges and local government officials have become increasingly sympathetic to community and environmental concerns when considering new housing developments. Zoning has become more restrictive. Bribery, they suggest, has become a less effective method for persuading officials to permit development. The increasing share of Americans who own homes has given homeowners more political clout, and homeowners have an incentive to limit the supply of new housing. Finally, rising educational levels and the lessons learned from other political battles, such as the civil rights movement, have made community members more adept at using courts and the press to battle development. Incumbent homeowners in the most desirable cities (such as San Francisco) gain the greatest home price appreciation from such supply restrictions because they own a scarce, valuable asset.

and their electromagnetic fields raises cancer risk. If no one is aware of this fact, then homes close to these power lines will not sell for a discount.[38] Or consider the evolution of home prices in Hong Kong in the wake of the severe acute respiratory syndrome (SARS) epidemic. One analyst found that apartment prices fell by an average of 3 percent in buildings housing SARS-affected tenants.[39] If prices return to their pre-SARS level, will this mean that the conditions that facilitated the spread of SARS have been corrected or simply that people have forgotten about this particular threat?[40]

Similarly, this approach can only account for local environmental goods that people care enough about to pay for, such as clean air, clean water, and the absence of hazardous waste sites. It cannot capture broader environmental problems, such as a city's contribution to the likelihood of climate change. This explains why looking at real estate prices can sometimes lead to counterintuitive conclusions about urban sustainability. Housing prices in southern cities such as Miami, for example, have risen steadily as the diffusion of cheap air conditioning has allowed households to enjoy its winter warmth without suffering too much from summer humidity.[41] But this improvement in quality of life has been purchased at the cost of a sharp increase in electricity usage. By the same token, the proliferation of green golf courses has helped increase demand for homes and consequently home prices in Las Vegas, despite the drain on water and land resources they represent.

Finally, if people lack the ability to "vote with their feet," real estate prices and migration patterns may offer a misleading picture of an area's desirability. This is true whether financial factors, legal restrictions, or mere sentiment limit individuals' ability to move in and out of a city. For

---

38. Conversely, the hedonic approach may overemphasize environmental threats that people think are real but that scientists believe are minor. Continuing with this earlier example, if home buyers greatly fear proximity to power lines, then homes located near them will sell for a deep price discount. If electromagnetic field exposure poses no real threats to households, then environmental scientists and the public might have opposite views on a given community's environmental quality.

39. Wong (2005).

40. A similar question is raised by a study of the impact of hurricanes on home prices in Pitt County, North Carolina. After examining data from 1992 to 2002, Okmyung Bin and Stephen Polasky found that the discount for a house located within the floodplain increased significantly after Hurricane Floyd hit the region in September 1999 (Bin and Polasky 2004). Does this mean that the risks associated with living there increased after 1999 or that people had better information?

41. Rappaport (2003).

example, many households have family living in the cold Northeast. If these familial ties cause Northeastern households to stay put, then a researcher who does not observe this variable might conclude that his sample families really like living in the Northeast while the truth may be that they tolerate the Northeast in order to live near their relatives.

Building a Green City Index

Each of these approaches to measuring urban environmental quality has strengths and weaknesses. For example, the footprint approach does the best job of judging a city's impact on the regional and global environment. But the public health approach is better at capturing the heterogeneous impact of environmental threats. In a heterogeneous society, policymakers often care about how the *average* person is doing whereas activists often demand that special attention be paid to the quality of life of the least well-off. The public health approach provides data that both camps can use. By contrast, the ecological footprint implicitly assumes that increased carbon dioxide production leaves all people equally worse off. Since cities differ with respect to their location and geography (compare New Orleans to Minneapolis), the residents of some northern cities will actually gain from climate change—at least insofar as average daily temperatures are concerned.

Similarly, despite the disadvantages discussed above, the compensating differentials approach has the advantage of accounting for local environmental risks whose consequences have yet to be realized. For example, imagine that the link between an environmental hazard and a particular disease is widely recognized, but the disease has a long incubation period. In this case current health data would indicate no impact, but homes would sell at a discount reflecting anticipated future costs.

In sum, not only does each of these methods provide a different piece of the picture, but the conclusions they lead to can often diverge. A city with a small ecological footprint may be strewn with disease-promoting waste; a city with plenty of green parks may be generating enormous quantities of carbon dioxide and other gases that make climate change more likely. The latter situation has become particularly common in the United States over the last thirty years. Many urban local environmental criteria are improving while overall resource consumption and carbon dioxide emissions from urban transportation have increased. What is the net effect of these trends for ranking cities?

To make precise statements about whether San Francisco, say, is "greener" than Houston, analysts must take a stand on the relative importance of local and global environmental indicators. To appreciate this point, suppose that based on objective data, Houston receives a grade of a "green A" according to local environmental criteria and a "green C" according to global environmental criteria, while San Francisco receives a "green B" in both cases. Which city has a higher "green grade point average"? If both sets of criteria receive equal weight, then the two cities have equal green indices. But if more importance is placed on local environmental criteria, then Houston will be ranked as the greener city. In other words, based on the same objective environmental data, two different people may rank cities differently depending on how they prioritize different threats.

To present the problem more formally, consider the following "green city index":

$$\begin{aligned}\text{Green city index} = \; & (b_1 * \text{environmental morbidity}) \\ & + (b_2 * \text{environmental mortality}) \\ & + (b_3 * \text{pollution avoidance expenditure}) \\ & + (b_4 * \text{local disamenities}) \\ & + (b_5 * \text{ecological footprint}),\end{aligned}$$

where b stands for index weights. (See discussion below on deriving these weights.) Cities with low scores on this index are green cities; those with high scores are brown.

In this equation environmental morbidity represents the number of sick days that the city's average resident experiences as a result of being exposed to urban pollution. Similarly, environmental mortality represents the additional risk of death the average resident of this city faces due to environmental threats. Pollution avoidance expenditure represents per capita protective investments, such as bottled water, which typically rise with pollution. Local disamenities represent factors, such as limited access to parks, that reduce urban environmental quality without necessarily increasing environmental morbidity or mortality. Finally, the ecological footprint can be represented by per capita carbon dioxide emissions, measured in tons.

Unfortunately, the necessary data are lacking to construct this index for cities in the United States or across the world. For example, there are no data on per capita carbon dioxide production for major cities across a set of years, nor are there data for the average environmental mortality risk associated with living in a given city. Throughout this book I try to address this problem by using what data are available to present new facts about the index's components. For example, in chapter 7, I use data from 2001 to estimate average annual gasoline consumption for many major U.S. cities. This can serve as an alternative method for comparing per capita ecological footprints across cities.

Nonetheless, going through the thought process involved in constructing a green city index helps highlight the subjective choices such an exercise involves. In particular, it forces us to confront the question of how we should choose the index weights b_1 through b_5. Of course, all efforts to create rankings based on multiple criteria run into the same problem. For example, to generate its annual report on the best places to live, *Money* magazine first collects data on city attributes, such as crime rates, commute times, and the availability of cultural options, and then assigns a weight to each indicator before combining the indicators into a single index. How should these weights be selected? The simplest approach is to assign each indicator equal weight, but this approach is often intuitively unappealing. An alternative is to poll the likely users of the index on their priorities. The Zagat dining guides adopt a more agnostic approach by simply providing separate scores for each criterion, such as food and service, for each restaurant without aggregating them into a single index.

While many approaches are possible, an economist would tackle this problem by relying on the notion of revealed preferences—that is, rather than guess the value that individuals attach to some good, the decisions people actually make would be used to tease out how much they think something is really worth. For example, as previously discussed, labor economists would advocate using a person's daily wage as a measure of the value of a lost day of work (b_1). They would also use risk premiums or similar data to estimate the value of a statistical life (b_2). Since pollution avoidance expenditures are already measured in dollars, they already have a marketplace value, and b_3 can simply be set to equal one. As for b_4, it can be estimated by applying the hedonic methods presented earlier in this chapter. Finally, what about b_5? What weight should we place on the importance of a city's ecological footprint? World Bank researchers have been wrestling with this problem because they are attempting to adjust national income accounts to account for natural capital depletion. These researchers have chosen to value carbon dioxide emissions at $20 per ton, based on estimates of the marginal damage to the world's environment caused by an extra ton of carbon dioxide emissions.[42] Thus they would set b_5 equal to $20.

Each of these choices can be challenged. Environmentalists may argue that the size of a city's ecological footprint should receive greater weight. Given that increases in morbidity and mortality are most likely to affect the poor, activists for social justice might prefer to increase the weights attached to these indicators and reduce the weight assigned to pollution avoidance expenditures, which typically increase with income. When it comes to local disamenities, analysts may disagree about how much influence they should have on the index, depending on whether freedom from the disamenity in question is seen more as a luxury or a basic need. As a result constructing a green city index that will satisfy everyone is probably an impossible task. However, by thinking in these terms, we can gain a better sense of what is meant when we say a city is green—and of how different yardsticks can be combined to test our intuitive notions of which cities are green and which are brown.

In this book I examine a variety of urban environmental indicators, ranging from water pollution to gasoline consumption. Through this process I seek to explain how urban quality of life evolves as urbanites grow richer and more numerous and as cities sprawl. In some cases the links between these indicators and day-to-day urban quality of life are obvious. If the local air is highly polluted, local quality of life declines. But in other cases the relationship is more subtle. As individuals people raise their own quality of life when they consume more gasoline since the resultant ease of mobility allows them to achieve their daily goals. But in aggregate, by promoting climate change, gasoline consumption can lower urban quality of life. Using all three yardsticks—those based on ecological, public health, and economic standards—makes it possible to capture these effects and paint a more comprehensive picture of the environmental consequences of urban growth.

---

42. Arrow and others (2004).

CHAPTER 3

The Urban Environmental Kuznets Curve

In 1940 New York City had more air pollution than in 1800; in 2000 it had less air pollution than in 1940. Today, an industrializing city like Bangkok features higher pollution levels than either a poor city like Accra or a rich city like Paris. Whether one compares the same city's pollution levels over time or pollution levels across cities today, an interesting pattern emerges. By many indicators environmental quality initially declines as poorer cities develop, but as growth continues, a turning point is eventually reached, and thereafter, environmental quality improves as income rises. This distinctive pattern is known as the environmental Kuznets curve (EKC).

Why is development both a foe and friend of the urban environment? Over the past fifteen years, a great deal of work in environmental economics has sought to document the complex relationship between measures of income and environmental quality. This literature focuses on how consumers, producers, and governments respond as economic development takes place.

The EKC Hypothesis

The environmental Kuznets curve hypothesis posits a nonmonotonic relationship between development and environmental quality. Figure 3-1 presents two examples of an EKC. The basic idea is that economic development initially increases pollution levels due to growth in the

Figure 3-1. *Two Examples of an Environmental Kuznets Curve*

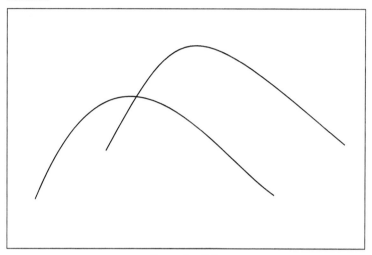

scale of production and consumption. For example, as people grow richer, they switch from bicycles to cars and live in larger homes that use more appliances. In time, however, development also triggers offsetting effects—notably by shifting consumption and production in greener directions and by giving policymakers the mandate and the resources to implement regulation that reduces pollution. For example, requiring car companies to produce vehicles with lower emissions per mile could lead to an overall decline in air pollution, even if rising affluence causes the total number of miles driven to increase. If the EKC hypothesis is true, trends in per capita income underestimate overall changes in well-being in rich cities but overestimate such trends in poorer cities.[1]

1. Nobel laureate Amartya Sen has conducted extensive research on the human development indicators (Anand and Sen 1994). This research attempts to rank quality of life across different countries. Intuitively, if a person lives in a nation with a moderate level of income per capita, but this nation also offers freedoms and other opportunities, then such a country is likely to score higher on Sen's index than one with a higher level of per capita income but ruled by a dictatorship. Returning to the EKC, if richer cities are experiencing environmental improvements, then because national income accounting does not reflect such nonmarket environmental indicators, the improvements in these indicators mean that GNP growth understates progress in well-being in such nations. In this sense the EKC research agenda

Some World Bank economists have argued that over time the EKC is shifting down and to the left.[2] If this is true, developing nations can expect to benefit in two ways. First, they are likely to reach the income turning point earlier in their development than nations have in the past. Second, they will suffer less environmental damage before reaching the turning point. Unfortunately, data limitations make it very difficult to test this hypothesis. A statistician intent on determining whether the EKC relationship is stable over time would need to collect city-level data on environmental quality and per capita income over a long period. While there are good data on different nations' ambient particulate levels in recent years, no such data exist for 1950, let alone for 1900 or 1850. Even the United States has only been monitoring ambient air pollution since the early 1970s. Similarly, the commonly used Global Environmental Monitoring System (GEMS) data set, which includes measures of total suspended particulates sampled at 149 monitoring stations in fifty-three cities across thirty countries, is only available for the years 1971 to 1992.[3]

These data limitations are unfortunate because the shape and location of the EKC have important welfare implications. As a World Bank team points out,

> The stakes in the environmental Kuznets curve debate are high. Per capita GDP in 1998 (in purchasing power parity dollars) was $1440 in the nations of sub-Saharan Africa, $2060 in India, $2407 in Indonesia, and $3051 in China. Since these societies are nowhere near the income range associated with maximum pollution on the conventional environmental Kuznets curve (typically $8000 per capita), a literal interpretation of the curve would imply substantial increases in pollution during the next few decades. Moreover, empirical research suggests that pollution costs are already quite high in these countries. For example, recent World Bank estimates of mortality and morbidity from urban air pollu-

---

can be viewed as an extension of "green accounting" research (Arrow and others 2004).

2. Dasgupta and others (2002).

3. The World Health Organization and United Nations sponsored the creation of this data set, which also includes measures of ambient sulfur dioxide sampled at 285 monitoring stations in 102 cities across 45 nations. It is often used to estimate ambient air pollution EKC curves. See Grossman and Krueger (1995); Harbaugh, Levinson, and Wilson (2002).

tion in India and China suggest annual losses in the range of 2–3 percent of GDP.[4]

Depending on the EKC's shape and location, the costs associated with further development could be much higher or lower than current estimates suggest.

Origins of the EKC

The EKC hypothesis originated in the early 1990s debate concerning the environmental consequences of the North American Free Trade Agreement (NAFTA), which was signed by the United States, Canada, and Mexico. Some analysts argued that NAFTA would degrade Mexico's environmental quality due to free trade's scale and composition effects. Trade liberalization would increase economic activity in Mexico and thereby generate more pollution. In addition, environmentalists feared that as trade opened up, relatively stringent U.S. regulation would push dirty production south of the border. This would increase the role that dirty industries played in Mexico's economy. Furthermore, the critics maintained, overall North American pollution would increase because the relative laxity of Mexican regulation would result in more pollution per unit of production than if the same production had taken place in the United States.

Two Princeton University economists, Gene Grossman and Alan Krueger, countered this critique by arguing that the environmentalists ignored trade liberalization's more benign effects. Free trade would lead to an increase in foreign direct investment in Mexico and the importation of cleaner technologies. In addition, by encouraging specialization, free trade would increase Mexico's per capita income. Since richer households typically demand greater environmental protection, the result could be more effective environmental regulation. As Grossman and Krueger wrote, "More stringent pollution standards and stricter enforcement of existing laws may be a natural political response to economic growth."[5] The Clinton administration endorsed this view. To quote Vice President

---

4. Dasgupta and others (2002).
5. Grossman and Krueger (1995). Enough time has now passed to allow researchers to examine NAFTA's effects. In *Free Trade and Beyond*, Kevin Gallagher provides one such evaluation (Gallagher 2004). In a review of this analysis, Richard Feinberg states that Gallagher "provides ammunition for both defenders and detractors of the North American Free Trade Agreement. In a conclusion consistent with

Albert Gore Jr., "Free trade will accelerate economic progress, making greater resources available for environmental protection."[6]

Grossman and Krueger tested their hypothesis with cross-national data on ambient sulfur dioxide and total suspended particulates from the GEMS database.[7] Based on these measurements, they estimated the following model:

Air pollution = controls + (b_1*GNP) + (b_2*GNP^2) + error term,

where b_1 and b_2 are regression coefficients that are estimated using ordinary least squares and where GNP represents a nation's real per capita gross national product. By including the term GNP^2 in this statistical model, researchers can test for whether there is a nonlinear relationship between pollution and national per capita income. The results of this regression showed b_1 to be positive and b_2 to be negative, yielding the hill-like shape of the EKC. Grossman and Krueger also indicated that the peak of the hill—or the income turning point—was located between $6,000 and $8,000 per capita, depending on which measure of pollution the researchers used.

Recently, William Harbaugh, Arik Levinson, and David Wilson reexamined these findings and concluded that they are not robust to minor changes in the econometric specification.[8] Based on their analysis, the available data do not definitely show or reject the existence of a cross-national EKC for particulates or sulfur dioxide. Nonetheless, many other researchers have identified environmental indicators that appear to follow the EKC pattern. In the next section, I discuss two such indicators that have particular relevance for cities.

---

other expert findings, Gallagher states that Mexico has not served as a pollution haven; there has been no 'race to the bottom.' Economic growth has continued to degrade Mexico's environment, yet he cannot isolate and therefore cannot credibly blame international trade and investment. At the same time, Gallagher finds that NAFTA has failed to halt the damage caused by growth to Mexico's air and water; its environmental institutions have generated some good pilot programs, but they lack the money and power to carry real bite. Mexico, for its part, has developed an elaborate set of environmental laws but not the political will or resources to enforce them. (With apparent calculation the government sharply increased the number of plant inspections just before NAFTA's ratification but decreased them precipitously soon thereafter.)" See Feinberg (2005).

6. Vice President Albert Gore, statement at the signing of the Uruguay Round Agreement, Marrakesh, March 1994.
7. Grossman and Krueger (1995).
8. Harbaugh, Levinson, and Wilson (2002).

Two Examples of the Urban EKC

Noise pollution is a leading example of the urban EKC. Almost all forms of urban economic activity generate noise, and as the scale of such activity increases, noise pollution rises. In a growing city, there will be more driving as motorcycles and cars replace bicycles, more manufacturing, and more construction activity. If antinoise laws are not enforced, then noise levels will grow. Eventually, however, richer consumers are likely to upgrade the quality of the products they buy or press successfully for noise regulation. At this point the EKC turning point will be reached.

Developments in air travel illustrate this effect. In many major cities, airports are major noise producers. As air travel increases at busy airports, it is certainly possible that noise levels around the flight path could sharply rise. Yet noise pollution has been falling near Chicago's O'Hare airport, despite growth in the number of flights. As one observer explains, "Airports are actually becoming significantly quieter over time. New aircraft are much quieter than older planes and the older aircraft are being retired. Indeed, a single model, the B72Q, which is being phased out by the major airlines, generated over 70 percent of the incidences of 'severe noise' at O'Hare in 2001. In addition, airports have become quieter as night flights are reduced."[9]

Richer cities build highway sound walls, enforce antihonking laws with stiff fines and higher probabilities of detection, and enforce zoning laws separating residential communities from noisy industrial areas. Richer urban residents also have additional self-protection options, such as purchasing thicker windows to block out unpleasant noise. All these factors are reflected in the downward slope of the EKC.

Similarly, recent research has found evidence of an EKC at the cross-national level for lead emissions.[10] This is important for public health because lead exposure has a number of negative impacts on child development. For example, high lead levels can lower IQs and contribute to psychological problems.

Leaded gasoline is a major contributor to urban lead exposure. Initially, as nations get richer, they consume more leaded gasoline. Richer urbanites are more likely to own vehicles and drive more miles. These

---

9. McMillen (2004).
10. Hilton and Levinson (1998).

scale effects raise ambient lead levels. To provide some evidence on the relationship between income growth and the scale of driving, I use cross-national data from the World Development Indicators database.[11] This data set provides information on per capita gross national product and vehicle ownership. Using data for 158 nations in 1996, I find that the income elasticity of vehicle ownership is 0.91. This means that a 10 percent increase in per capita income increases vehicle ownership rates by a little more than 9 percent. A dramatic example of the impact of income growth on vehicle ownership rates is provided by Beijing, where car ownership has been growing by 15 to 20 percent per year. Between 1997 and 2003, the total number of vehicles in the city doubled from one to two million.[12]

In the short to medium term, rising vehicle use can push airborne lead to dangerous levels. But richer nations are more likely to enact tighter standards regulating the permissible lead content of gasoline. Using data for 105 nations in the late 1990s, Per Fredriksson and colleagues found that all else being equal, a doubling of a nation's per capita gross domestic product reduced the permissible lead content of gasoline by 5.72 grams per liter.[13] Consequently, although richer nations consume more gasoline per capita, lead emissions per gallon of gasoline are relatively low. As the quality effect dominates the scale effect, lead levels fall. This dynamic causes urban lead emissions to follow the EKC pattern, with the turning point coming at roughly $11,000 in per capita income, based on a study that examined data for 48 nations from 1972 to 1992.[14]

While it is good news that such a turning point exists, it must be noted that in the year 2000, per capita income fell short of $11,000 in roughly 80 percent of the world's nations. This fact highlights the importance of designing institutions and creating incentives to shift the EKC down and in. It should also be noted that in terms of achieving urban sustainability, one would hope that urban pollution declines

---

11. Country profiles are drawn from World Bank, "World Development Indicators 2006—Key Development Data and Statistics" (worldbank.org [May 2006]). This online source is the World Bank's primary database for cross-country comparable development data.

12. Huang (2004).

13. The sample average across nations is 47.32 grams per liter. Fredriksson and others (2005).

14. Hilton and Levinson (1998).

sharply as a nation's GDP rises above the EKC turning point. If a city's pollution levels decline just slightly when per capita income passes the turning point, then it is difficult to argue that economic growth is increasing urban sustainability.

Shifting the EKC

The Kuznets curve is not a law of physics. The damage imposed by some environmental threats, such as waterborne diseases or natural disasters, appears to decline monotonically as per capita income grows. Cholera epidemics, for example, become less likely as countries become richer.[15] Similarly, while richer nations experience roughly the same number of earthquakes as poorer nations, richer nations suffer fewer deaths from similar earthquakes.[16] Between 1980 and 2002, India experienced fourteen major earthquakes that killed a total of 32,117 people while the United States experienced eighteen major earthquakes that killed only 143 people. Worldwide, controlling for national geography, population, and each earthquake's Richter scale reading, a 10 percent increase in per capita GDP decreases national earthquake deaths by 5.3 percent.[17]

Moreover, even when the EKC pattern is observed, a range of market and nonmarket forces can cause both the shape and the location of the curve to differ across environmental indicators, nations, and time.

---

15. In both this and the following case, the EKC may initially appear to hold. Cholera epidemics are likely to increase as nations first industrialize and then later decline. However, if one controls for population density, the bell-shaped relationship between income and cholera deaths will most likely disappear.

16. Rising incomes lead people to purchase higher-quality structures that might be better able to withstand natural shocks. Richer people will demand homes located in safer communities as well as homes that are built out of stronger, more durable materials. Once the shock has taken place, death counts can be higher if the nation does not have access to good medical care, emergency treatment, and crisis management (Athey and Stern 2002). Government regulation also plays a role in protecting the populace in richer nations. Richer nations will be able to invest in and enforce zoning and building codes. Building codes internalize the externalities of structural soundness into a building. In the absence of such regulation, building owners are unlikely to internalize the fact that their action improves the quality of life of people in the immediate vicinity (Cohen and Noll 1981).

17. Kahn (2005). For national annual data on earthquakes from 1980 to 2002, see Center for Research on the Epidemiology of Disasters, "EM-DAT: The International Disaster Database" (www.em-dat.net/index.htm [March 2006]). See also International Federation of Red Cross and Red Crescent Societies (2002).

Broadly speaking, these factors can have two types of effects on the EKC. First, they can move the curve down, which reduces the amount of environmental hazard associated with every income level but has no effect on the income turning point. Second, they can shift it to the left, which moves the turning point. Combining these two effects produces the optimistic scenario shown in figure 3-1.

Market Forces: Prices and Technology

Prices are the most basic factor affecting the shape and location of the EKC. Higher prices, notably for energy, retard economic development. According to James Hamilton, seven of the eight recessions in the United States between the years 1945 and 1980 were preceded by a dramatic increase in the price of crude oil.[18] However, a benefit of higher energy prices is that they create incentives to reduce consumption. In addition, price shocks can have the unintended consequence of "greening" consumption and production techniques. For example, if drivers believe that energy price increases, such as those imposed by OPEC in the 1970s, will persist, then they will reduce their purchases of gas-guzzling sport-utility vehicles. Not only will this reduce carbon dioxide production, but it will also encourage vehicle makers to invest in more fuel-efficient technologies, such as hybrid cars.

This example illustrates the induced innovation hypothesis, which argues that higher expected energy prices create incentives for producers to develop and sell products that economize on energy consumption.[19] Recent research supports this hypothesis by suggesting that firms redirect their research and development budgets in response to changing market fundamentals. For example, the number of energy patents granted increases soon after the price of energy rises.[20] Similarly, studies indicate that rising energy prices have induced energy-saving innovation in individual sectors, such as air conditioning equipment. According to one study, rising energy prices account for roughly a quarter to a half of the improvements in mean energy efficiency in air conditioners over the last two decades.[21]

---

18. Hamilton (1983).
19. Jaffe, Newell, and Stavins (2002); Moomaw and Unruh (1997).
20. Popp (2002).
21. Newell, Jaffe, and Stavins (1999).

Greener technologies play two roles in helping to explain the EKC. First, to the extent that demand for such technologies is related to income, the growing availability of alternatives like hybrid cars can help account for the curve's distinctive shape. Second, changes in the availability of such technologies can help shift the EKC. For example, cities in the developing world today may be able to meet the growing demand for air conditioning at lower environmental cost than U.S. cities thirty or more years ago. Similarly, cities with well-designed mass transit systems are likely to find that growth affects local environmental quality less dramatically than cities whose residents are more heavily dependent on cars. Such technological advances reduce the environmental impact of economic growth.

Nonmarket Forces: Government's Role

Like green technologies, government action also helps explain both the shape and location of the EKC. Some economists argue that as income rises, the typical voter becomes more willing to support spending on regulation.[22] A related hypothesis suggests that a nation is more likely to enact environmental regulation when its economy is growing and income inequality is falling. Under these conditions voters are more likely to agree on policy priorities, and the government has the resources necessary to pursue environmental goals. These types of arguments help account for the shape of the EKC.

In addition, political action can shift the EKC turning point or help move the curve down. One unexplored but plausible channel for such influence consists of firm expectations. If companies expect "green" politicians to remain in office over an extended period, they will have stronger incentives to invest in energy-saving and pollution-reducing technologies. By contrast, a "pro-business" government is unlikely to stimulate downward and inward movement in the EKC.

Even more important, factors or events that help trigger pro-environmental activism can have a significant impact on the location of the EKC turning point. For example, better information about the effects of environmental hazards could cause demand for regulation to be voiced at a lower level of per capita income, moving the curve in. Such information is often provided by dramatic events, such as the 1969 fire on

---

22. Selden and Song (1995).

Figure 3-2. *Pollution Level and Death Rate, London, December 1–15, 1952*

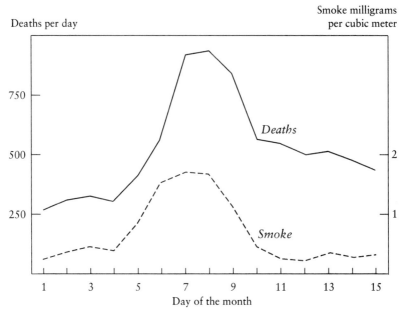

Source: Met Office, "The Great Smog of 1952" (www.metoffice.com/education/secondary/students/smog.html [November 2005]). © Reprinted with permission, Crown copyright 2006; published by the Met Office.

Ohio's Cuyahoga River or the 1989 *Exxon Valdez* oil spill, that wake up a complacent electorate to lobby for greater environmental protection. For example, horrific pollution levels helped trigger major changes in air pollution policy in London in the 1950s. Figure 3-2 presents time series data documenting the dramatic increase in London's pollution levels in 1952 and the subsequent rise in the death rate. The silver lining of this tragedy was the fact that it prompted the passage of far more stringent air pollution regulation. Following the Great Smog of 1952, the British government passed the City of London (Various Powers) Act of 1954 and the Clean Air Acts of 1956 and 1968. These laws banned emissions of black smoke and decreed that residents of urban areas and operators of factories must convert to smokeless fuels.[23]

23. Met Office, "The Great Smog of 1952" (www.metoffice.com/education/secondary/students/smog.html [November 2005]).

Figure 3-3. *Articles Mentioning Pollution*, New York Times, 1979–2001

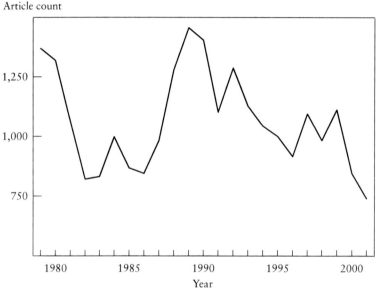

Source: Author's calculations based on database from ProQuest Information and Learning, "ProQuest Historical Newspapers" (www.proquest.com/products_pq/descriptions/pq-hist-news.shtml [May 2006]).

While the consequences of London's Great Smog were immediately obvious to anyone living in that city at the time, in other cases people may only be aware of environmental challenges if they are publicized in media outlets, such as newspapers and radio and television programs. As a result the existence of a free and competitive media establishment can help bring about environmental regulation at an earlier stage in a nation's development. Figure 3-3 presents annual data from the *New York Times* archives documenting yearly trends in the number of articles in this influential newspaper that mention pollution.[24] There is clear evidence that there were many more articles written about the environ-

---

24. *New York Times* articles were counted in monthly increments following the environmental shock. The articles were found using the ProQuest Historical Newspapers database for the *New York Times*, 1851–2003. Keyword searches were used to find the different environmental shocks, such as "Chernobyl." See ProQuest Information and Learning, "ProQuest Historical Newspapers" (www.proquest.com/products_pq/descriptions/pq-hist-news.shtml [May 2006]).

ment in the late 1980s than in more recent years. Such intensive coverage can potentially shift the EKC by transforming local tragedies into salient events that trigger regulation on a national scale.

Consider the example of Love Canal, New York. Throughout the 1940s and 1950s, more than 21,000 tons of chemical wastes were dumped there. In the 1970s Love Canal residents began to complain of health problems, including high rates of cancer, birth defects, miscarriages, and skin ailments. The Environmental Protection Agency claims that 56 percent of the children born in Love Canal between 1974 and 1978 had birth defects. The Love Canal disaster helped galvanize support for programs to address the legacy of industrial waste, and this political pressure led to the creation of the Superfund Program in 1980.[25] As Timur Kuran and Cass Sunstein described this process,

> A kind of cascade effect occurred, and hence in the period between August and October 1978, the national news was saturated with stories of the risks to citizens near Love Canal. The publicity continued in 1979 and 1980, the crucial years for Superfund's enactment. There can be no doubt that the Love Canal publicity was pivotal to the law's passage in 1980. In that year, *Time* magazine made the topic a cover story, and network documentaries followed suit. Polls showed that eighty percent of Americans favored prompt federal action to identify and clean up potentially hazardous abandoned waste sites. Congress responded quickly with the new statute.[26]

Another example of the importance of information has its roots in the Union Carbide plant disaster in Bhopal, India, which prompted the creation of the U.S Toxic Release Inventory. Through this inventory local communities are now supplied with information from local manufacturing plants concerning which chemicals and in what quantities these plants are releasing into the local environment. This information is widely publicized through the media, and its release creates a "day of shame" for the largest polluters.

Environmentalists recognize the power that salient events can have in motivating voters to take and support costly action.[27] Without salient

---

25. Greenstone and Gallagher (2005).
26. Kuran and Sunstein (1999).
27. Disasters are not the only source of such events. Bestselling books, such as *Silent Spring* (Carson 1962) and *The Population Bomb* (Ehrlich 1968), can be

events it becomes much harder to mobilize a coalition to enact green regulation. Therefore distant, ambiguous environmental issues, which are unlikely to grab the public's attention, are often the most difficult to resolve. A savvy politician who wants to be reelected will not lead the fight on global warming when he can gain more from winning the war on crime or improving school quality.

"Lulling" events make the environmentalists' job even harder. If people are lulled into believing that environmental problems are improving over time, then they will be less likely to support an activist environmental regulatory agenda. Fear of this effect may explain why books that are optimistic about ongoing environmental trends often provoke such strong controversy. For example, Bjørn Lomborg's book *The Skeptical Environmentalist* was a media sensation, generating repeated headlines in the *New York Times*, despite its technical detail, graphs, and data analysis. The media was captivated by the fact that an ex-Greenpeace member was optimistic about global sustainability trends, and environmentalists quickly mobilized to rebut his conclusions.[28]

Arguments against the EKC Hypothesis

Environmentalists have voiced several criticisms of the EKC hypothesis. Four of the leading concerns focus on irreversibilities, short-run environmental degradation, the role of pollution havens, and the challenges posed by cross-border externalities.

Earlier, I discussed two examples of urban pollution—noise and lead emissions—that are "reversible" environmental indicators. If noise or air pollution increase one year, it is quite possible to reverse these trends in the next year. Embedded in the EKC hypothesis is the assumption that past environmental damage can be undone in the future. Clearly,

---

thought of as catalytic cultural events. Recently, some Hollywood filmmakers have attempted to create salient events. In his movie *A.I.*, Steven Spielberg displays a flooded New York City. In this future scenario, only the Statue of Liberty's torch is above water.

28. See Lomborg (2001). This book actually served as a catalyst for me to write this book. I was intrigued by his empirical evidence documenting long-run trends for many environmental indicators, but I thought that a weakness of the book was that it did not carefully study polluters' incentives to pollute. His book also devoted little attention to the government's role in mitigating pollution externalities and the conditions under which one could be optimistic that government would play this role.

for certain pieces of the environment this reasoning is faulty. Consider species extinction: if a species becomes extinct, no government action can bring it back.

A second valid concern with the EKC is that even if the curves shown in figure 3-1 represent the true dynamic path for a city's environmental quality, there are many cities in the developing world that currently lie far to the left of the hypothetical turning point. In these cities rapid population growth and rising per capita income combine to cause significant environmental degradation as the scale of consumption and production grows. In many cases environmental quality is likely to fall for decades—or even longer—before the income turning point is reached. Consider a city where per capita income is $3,000 with a growth rate of 2 percent a year. Under these conditions it will take thirty-six years for this city to achieve a per capita income of $6,000, which may still be below the income turning point for amelioration of certain environmental hazards.

Pollution Havens Hypothesis

Some critics have also used the pollution havens hypothesis to question the significance of the EKC. This hypothesis relies on the fact that poorer nations tend to have less stringent environmental regulation than richer nations. Such lax regulation may act as a magnet for dirty industries. If rising incomes in the developed world lead to more regulation, which causes more dirty production to move to developing countries, then the EKC may overstate the net environmental benefits of income growth. While environmental quality is improving in countries to the right of the EKC turning point, as the hypothesis predicts, these gains are matched by declines in other parts of the world.[29]

The evidence for the pollution havens hypothesis is mixed. Some research investigating trade patterns and foreign direct investment supports the hypothesis, particularly where footloose, lightweight industries, such as jewelry or office and computing machines, are con-

---

29. For example, it is quite intuitive that one might observe an EKC for garbage in richer cities that are physically close to poorer cities. As the richer city develops, it could pay the poorer city to take its trash. Such trades could take place between nearby cities in western and eastern Europe, for example. For the richer city, this trade would yield a curve like those shown in figure 3-1, but *global* garbage production would not be declining. I will return to this point in chapter 5.

cerned.[30] But for most industries pollution control costs are not a major determinant of relocation.[31] Survey evidence shows that environmental regulatory costs rank low among the factors that influence firm locational choice.[32]

Of course, dirty producers may have good reasons to migrate to less regulated poorer nations, even if escaping regulation is not their primary goal. Labor economists have found ample evidence that "dirtier" jobs (whether measured in job safety, long hours, or pollution exposure) must pay higher wages to attract workers.[33] As real wages in the United States rise, so does the wage premium that workers require for performing such onerous tasks.[34] In response, firms may choose to move their plants overseas, assuming that the resulting cut in production costs is not swamped by an increase in transportation and other logistics costs.

But even this weaker version of the pollution havens hypothesis is not borne out by the evidence. Since the 1950s the pollution content of U.S. manufacturing imports has fallen relative to the pollution content of domestic manufacturing production.[35] Using data on annual production, imports, and exports for 435 manufacturing industries ranging from gum to steel, I study time trends in the pollution content of manufacturing trade. Since energy consumption is positively correlated with a number of environmental problems such as air pollution, water pollution, and greenhouse gas production, I construct each industry's annual energy consumption per dollar of value added and use that as an indicator of the pollution content of trade.[36]

Figure 3-4 shows how the average energy content of U.S. domestic production, imports, and exports has changed from 1958 to 1994. For each of the three lines, average energy content is calculated as

$$\text{Average energy content in year } t = \Sigma\ s(jt)^* X(j,t).$$

In this formula $s(jt)$ represents the share s of economic activity in industry j in year t. In any given year, the shares sum to one. $X(j,t)$ represents

30. Ederington and Minier (2003); Ederington, Levinson, and Minier (2005); Kahn (2003b).
31. Dasgupta and others (2002).
32. Panayotou (2000).
33. Hammitt, Liu, and Liu (2000).
34. Costa and Kahn (2004).
35. Kahn (2003b).
36. The data source is the NBER Manufacturing Productivity Database (Bartelsman and Gray 1996).

Figure 3-4. *Average Energy Content of U.S. Manufacturing, 1958–94*
Percent of dollar value added

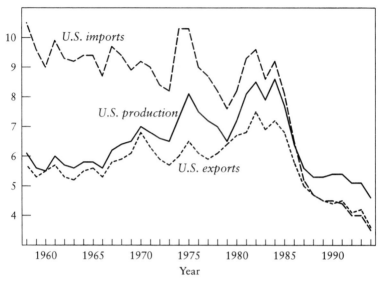

Source: Author's calculations based on data from Bartelsman and Gray (1996) and Feenstra (1996).

the pollution index X for industry j in year t. In 1958 the average energy content of U.S. imports was almost twice as high as that of exports and domestic production. But between 1958 and 1994, the average energy content of imports declined much more than the average energy content of domestic production and exports. This is suggestive evidence that production did not move to pollution havens, even after U.S. environmental regulations became increasingly restrictive, starting in the early 1970s.[37]

However, domestic pollution havens may substitute for international ones. Within the United States, for example, there are cross-state differences in the intensity of environmental and labor regulation. Some states, such as California, are known as regulatory leaders, while other states,

---

37. Copeland and Taylor (2004) argue that pollution-intensive industries also tend to be capital intensive. The factor endowment hypothesis predicts that nations endowed with high levels of capital relative to labor will be more likely to be dirty goods exporters. If this hypothesis is correct, then a rich nation such as the United States, despite stringent environmental regulation, may continue to produce dirty goods.

such as Texas, have the opposite reputation.[38] More broadly, right-to-work states, which ban union shops (workplaces in which all employees are required to join the union), also tend to have relatively lax environmental regulation. Based on data reported at the League of Conservation Voters web site, I calculate that in 2002 the average congressional representative from right-to-work states voted the pro-environment position 26 percent of the time versus 56 percent for representatives from non–right-to-work states.[39] A growing literature shows that such differences encourage production to move to less regulated areas.[40] Many manufacturers are likely to prefer this option to moving production to a developing country, where they would face exchange rate risk, higher transportation costs to U.S consumers, and potential political instability.

Cross-Border Externalities and the EKC

The final and perhaps most important criticism of the EKC hypothesis is that it is far less likely to hold for cases such as acid rain, where some of the costs of pollution are "externalized" onto neighbors. If India's electric utilities create sulfur dioxide that floats off to South Korea, then India has little incentive to enact regulations that could potentially slow economic growth. Similarly, officials in China's Guangdong province have little interest in limiting the manufacturing activity that has led to a sharp rise in Hong Kong's particulate levels.[41] As a result cross-boundary externalities can impose large environmental costs.

Greenhouse gas production represents the ultimate cross-border externality. When a nation produces more greenhouse gases, all nations face a higher risk from global climate change. However, unlike increases in noise or air pollution, an increase in the stock of greenhouse gases has no immediate impact on households. Consequently, in the absence of some salient event, such as an extreme summer heat wave, voters are unlikely to support costly measures to mitigate this externality. I will return to this point in chapter 8.

---

38. Fredriksson and Millimet (2002); Berman and Bui (2001); Levinson (2001).
39. League of Conservation Voters, "2002 National Environmental Scorecard" (www.lcv.org/images/client/pdfs/scorecard02final.pdf [May 2006]).
40. Becker and Henderson (2000); Greenstone (2002); Henderson (1996); Holmes (1998); Kahn (1997); Keller and Levinson (2002); Levinson (1996).
41. Keith Bradsher, "Rosy, Pink Cloud, Packed with Pollution," *New York Times*, September 10, 2002, p. A10.

Conclusion: When Does the EKC Appear?

The environmental Kuznets curve is a parsimonious model of how environmental quality evolves in a growing market economy. Many environmental economists have been interested in testing its predictions about changes in environmental quality across nations and over time. The result has been a large and engaging empirical literature. However, the EKC has also given rise to some potentially dangerous misconceptions, as Jagdish Bhagwati observes in comments on some of the many indicators that have been found to follow the EKC:

> The only value of these examples is in their refutation of the simplistic notion that pollution will rise with income. They should not be used to argue that growth will automatically take care of pollution regardless of environmental policy. Grossman and Krueger told me that their finding of the bell-shaped curve had led to a huge demand for offprints of their article from anti-environmentalists who wanted to say that "natural forces" would take care of environmental degradation and that environmental regulation was unnecessary; the economists were somewhat aghast at this erroneous, ideological interpretation of their research findings.[42]

The existence of the EKC hinges on several critical political and economic factors. Together, these will determine whether or not the trajectory of the green city index (as defined in chapter 2) will approximate a bell-shaped curve. First, there must be markets for the goods (or bads) in question. For types of natural capital, such as minerals and gasoline, where formal markets exist, economic development can actually shrink a society's ecological footprint in the long run. Rising incomes increase demand for such natural resources, and the prices of such resources typically rise as a result. This in turn triggers the induced innovation discussed earlier in this chapter. But in the absence of markets, consumers and firms do not face the right incentives to economize on activities that create pollution. Greenhouse gases, such as carbon dioxide, are a leading example of environmental commodities for which markets do not currently exist.

Second, prices in these markets must not be distorted as a result of political pressure. Researchers have noted that for many commodities,

---

42. Bhagwati (2004), p. 145.

people continue to discover new resource "needs." For example, Michael Hanemann has described how the need for water has grown: "As time passed, many other uses were found—tubs for bathing, water borne sanitary waste disposal, outdoor landscape and garden watering, automatic clothes washers, swimming pools, automatic dish washers, car washing, garbage disposal, indoor evaporative cooling, hot-tubs, lawn sprinklers, etc. The result has been a constantly rising trajectory of per capita household water use."[43] If political pressure causes water prices to be set artificially low, a city's or nation's water consumption will rise dramatically as its population and per capita income grow. More rational pricing, on the other hand, will promote conservation and keep consumption growth in check. As the Los Angeles Department of Water Power proudly reported in its *Urban Water Management Plan Update 2002–2003,* "Conservation continues to play an important part in keeping the city's water use equivalent to levels seen 20 years ago."[44]

Finally, the manifestation of an EKC hinges crucially on effective governance, which in turn requires broad access to information. Even if rising incomes lead to greater interest in environmental regulation, voters must still give voice to this interest, and policymakers must respond by adopting and implementing effective regulation. Otherwise, the EKC will not be observed. This issue will be taken up in chapter 5.

---

43. Hanemann (2005).
44. See Department of Water Resources, "Chapter 22: Urban Water Use Efficiency," *California Water Plan Update 2005. Volume 2: Resource Management Strategies* (www.waterplan.water.ca.gov/docs/cwpu2005/vol2/v2ch22.pdf [March 2006]).

CHAPTER 4

Income Growth and the Urban Environment: The Role of the Market

Market forces play a fundamental role in shaping the urban environmental Kuznets curve. Rising income levels lead to changes in the urban economy's consumption and production patterns that have the unintended benefit of greening the city. Most important, people in richer cities are more likely to consume higher-quality products and to work in the service sector. These behavioral changes help offset the pollution-causing effects of increasing scale and put the economy on the downward slope of the EKC.

Greening Urban Consumption

A millionaire could afford to purchase ten times as much of the same goods as a person who makes $100,000 a year. Yet consumer purchase data do not suggest that richer people simply scale up the quantity of their consumption. Instead, richer people purchase more quality. In some cases, this quality upgrading will have little environmental impact. For example, there are few environmental benefits offered by a richer person choosing to buy steak rather than eating at McDonald's. But in other cases richer households choose higher-quality products that intentionally or unintentionally have environmental benefits.

The Demand for Green

Two factors help explain why wealthier urbanites are more likely to demand green products and services. The first is simply the fact that they have higher incomes. Buying green—whether it takes the form of organic produce or hybrid cars—is often more expensive than buying brown. But higher-income city dwellers are also more likely to buy green because they typically have more education.

Education gives people the tools they need to access and process information about how environmental hazards affect their own well-being, as well as that of the planet. Consequently, rising educational attainment tends to increase a person's demand for a clean environment. This often translates into greater support for environmental regulation (see chapter 5). In addition, it can have a significant impact on individuals' private decisions. People with more education may be more patient and willing to make long-run investments than less educated people.[1] Often, they are less likely to smoke and more likely to eat a healthy diet. They are also more likely to seek out products and services that deliver greater environmental benefits.

Within the home, for example, richer households will invest more in mitigating hazards such as exposure to lead. These actions can have public as well as private benefits. By encapsulating lead-containing paint (which has been banned in the United States since 1977) or making sure that surfaces painted with lead-containing paint are in good condition, homeowners do more than protect their families' health. They also contribute to local environmental quality by reducing the likelihood that flaking paint will release lead into the air, soil, or water.

Another important example of the greening effect of income can be seen in vehicle purchases. While richer people drive more than poorer people, they also tend to drive newer vehicles. Newer vehicles are equipped with better emissions control technology. As a result they generate fewer emissions per mile. If newer vehicles are, say, 50 percent cleaner than old ones, richer people may be responsible for less driving-related pollution than their poorer counterparts, even if they drive 25 percent more miles. In many cases this benefit is an unintended consequence of the desire for a new car, but it can still have a significant impact on environmental quality.

---

1. Becker and Mulligan (1997).

The data set from California's random roadside emissions test program provides an opportunity to estimate the size of this quality effect. These data were collected by the Bureau of Automotive Repair, which conducted emissions tests on 24,615 vehicles between February 1997 and October 1999 by pulling cars over at random in "Enhanced Areas" around the state.[2] The data set provides detailed information on each vehicle's emissions of hydrocarbons, carbon monoxide, and oxides of nitrogen, and on the vehicle itself, including such variables as vehicle type, model year, mileage, make, and weight. A simple comparison of vehicle emissions by model year highlights how much cleaner newer cars are: the average vehicle built in 1996 emits 91 percent less hydrocarbons than the average vehicle built in 1986.

What do these data say about the relationship between income and adoption of greener products? Table 4-1 reports mean vehicle emissions and attributes for an entire sample of randomly tested vehicles in California and for four income categories based on percentiles of the California zip code average household income distribution.[3] The low-income category is derived from those vehicles registered in zip codes with average incomes below $43,037 (the 25th percentile of the distribution); the middle-income category is based on registrations from zip codes with average incomes between $43,038 and $72,026 (75th percentile); the high-income category is derived from vehicles registered in zip codes with average incomes between $72,027 and $120,998 (95th percentile); and the very high income classification is based on registration zip codes where the average income lies above the 95th percentile of the distribution. As shown in Table 4-1, there is a sharp monotonic relationship between average income by zip code and vehicle emissions per mile. The average vehicle in the poorest zip codes emits 54 percent more oxides of nitrogen, 144 percent more hydrocarbons, and 122 percent more carbon monoxide per mile of driving than the average vehicle in the richest zip codes.[4] This result is particularly striking because the average vehicle in the poorest zip codes is only two years older than the average vehicle in the richest ones.

2. Kahn and Schwartz (2006).
3. The measure of income is average household income from the 2000 Census of Population and Housing for the zip code of the vehicle's registration. See Missouri Census Data Center, "Census 2000: Data Products, Information and Activities" (mcdc2.missouri.edu/census2000/ [May 2006]).
4. Kahn and Schwartz (2006).

Table 4-1. *Mean Vehicle Emissions and Attributes by House Income Level, California, 1997–99*
Units as indicated

Emissions and attributes	Income level[a]				
	All	Low	Middle	High	Very high
Hydrocarbons (parts per million [ppm])	107.894	136.846	111.990	76.589	56.060
Carbon monoxide (percent of exhaust)	0.770	0.997	0.792	0.543	0.447
Nitrogen oxide (ppm)	697.947	828.380	705.613	579.091	535.874
Model year (average)	1986	1985	1986	1987	1987
Light truck (percent of vehicles sampled)	30.2	30.7	30.2	30.0	26.4
Average household income (year 2000 dollars)	60,615.12	40,698.62	55,007.50	84,609.16	142,861.60
Sample size	25,119	4,626	14,689	5,199	605

Source: Author's calculations based on data from California's 1997 to 1999 random roadside emissions test program (Kahn and Schwartz 2006).

a. See text for explanation of income categories.

Do households' annual vehicle emissions also fall as incomes rise? To answer this question requires information detailing how much households drive per year. The 2001 National Household Travel Survey is a U.S. Department of Transportation effort sponsored by the Bureau of Transportation Statistics and the Federal Highway Administration. It collects data on both long-distance and local travel by the American public. This data set provides information on household income and household annual miles driven.[5] This information can be used to predict household miles driven as a function of household income.[6] Table 4-2 shows that richer households drive more than poorer households, but for middle-class households annual mileage does not rise sharply with income. Table 4-2 also reports predicted hydrocarbon emissions per mile for each of the income categories.[7] Multiplying predicted emissions

5. See U.S. Department of Transportation, "2001 National Household Travel Survey" (nhts.ornl.gov/2001/html_files/download_directory.shtml [May 2006]).

6. Formally, I estimate a multivariate regression where the dependent variable is annual miles driven and the explanatory variables are a four-dimensional polynomial of household income.

7. Formally, I estimate a multivariate regression where the dependent variable is vehicle emissions per mile of driving and the explanatory variables are a four-dimensional polynomial of household income.

Table 4-2. *National Household Level EKC Data on Predicted Hydrocarbon Emissions per Mile, Annual Miles Driven, and Total Hydrocarbon Emissions, 2001*
Units as indicated

Income (thousands of 2001 dollars)	Hydrocarbons per mile (grams)	Miles driven per year	Total annual hydrocarbon emissions (grams)	Income (thousands of 2001 dollars)	Hydrocarbons per mile (grams)	Miles driven per year	Total annual hydrocarbon emissions (grams)
0	5.394	2,190.440	11,815.440	46	2.381	18,204.770	43,343.370
1	5.292	2,580.692	13,657.990	47	2.345	18,468.410	43,311.760
2	5.193	2,971.288	15,428.770	48	2.310	18,727.290	43,268.040
3	5.095	3,362.039	17,128.930	49	2.277	18,981.400	43,213.030
4	4.999	3,752.760	18,759.670	50	2.244	19,230.700	43,147.510
5	4.905	4,143.272	20,322.250	51	2.212	19,475.170	43,072.240
6	4.813	4,533.396	21,817.980	52	2.180	19,714.790	42,987.950
7	4.722	4,922.957	23,248.180	53	2.150	19,949.540	42,895.320
8	4.634	5,311.786	24,614.230	54	2.121	20,179.410	42,795.020
9	4.547	5,699.714	25,917.540	55	2.092	20,404.400	42,687.710
10	4.462	6,086.580	27,159.540	56	2.064	20,624.490	42,573.990
11	4.379	6,472.220	28,341.660	57	2.037	20,839.680	42,454.430
12	4.297	6,856.479	29,465.400	58	2.011	21,049.970	42,329.600
13	4.218	7,239.203	30,532.240	59	1.985	21,255.370	42,200.040
14	4.139	7,620.242	31,543.670	60	1.961	21,455.890	42,066.230
15	4.063	7,999.450	32,501.210	61	1.937	21,651.530	41,928.670
16	3.988	8,376.682	33,406.360	62	1.913	21,842.310	41,787.800
17	3.915	8,751.799	34,260.660	63	1.890	22,028.250	41,644.060
18	3.843	9,124.665	35,065.610	64	1.868	22,209.370	41,497.860
19	3.773	9,495.146	35,822.730	65	1.847	22,385.690	41,349.570
20	3.704	9,863.113	36,533.540	66	1.826	22,557.240	41,199.570

21	3.637	10,228.440	37,199.530	67	1.806	22,724.060	41,048.180
22	3.571	10,591.000	37,822.200	68	1.787	22,886.170	40,895.750
23	3.507	10,950.690	38,403.050	69	1.768	23,043.630	40,742.540
24	3.444	11,307.370	38,943.540	70	1.750	23,196.460	40,588.870
25	3.383	11,660.940	39,445.130	71	1.732	23,344.710	40,434.980
26	3.323	12,011.300	39,909.250	72	1.715	23,488.440	40,281.130
27	3.264	12,358.330	40,337.340	73	1.698	23,627.700	40,127.530
28	3.207	12,701.930	40,730.790	74	1.682	23,762.530	39,974.410
29	3.151	13,042.000	41,090.980	75	1.667	23,893.010	39,821.950
30	3.096	13,378.460	41,419.290	76	1.652	24,019.190	39,670.340
31	3.043	13,711.200	41,717.020	77	1.637	24,141.130	39,519.750
32	2.990	14,040.140	41,985.520	78	1.623	24,258.910	39,370.340
33	2.939	14,365.190	42,226.050	79	1.609	24,372.610	39,222.240
34	2.890	14,686.280	42,439.880	80	1.596	24,482.290	39,075.580
35	2.841	15,003.320	42,628.250	81	1.583	24,588.030	38,930.490
36	2.794	15,316.230	42,792.350	82	1.571	24,689.920	38,787.070
37	2.748	15,624.960	42,933.370	83	1.559	24,788.050	38,645.420
38	2.703	15,929.420	43,052.450	84	1.547	24,882.510	38,505.630
39	2.659	16,229.570	43,150.710	85	1.536	24,973.380	38,367.790
40	2.616	16,525.320	43,229.250	86	1.526	25,060.770	38,231.970
41	2.574	16,816.630	43,289.120	87	1.515	25,144.770	38,098.240
42	2.533	17,103.450	43,331.350	88	1.505	25,225.490	37,966.660
43	2.494	17,385.710	43,356.930	89	1.495	25,303.040	37,837.280
44	2.455	17,663.390	43,366.840	90	1.486	25,377.530	37,710.170
45	2.418	17,936.420	43,362.020				

Source: Author's calculations based on U.S. Department of Transportation, "2001 National Household Travel Survey" (nhts.ornl.gov/2001/html_files/download_directory.shtml [May 2006]) and 1997–99 California random roadside emissions test data.

Figure 4-1. *Nationwide Household EKC Curve for Predicted Annual Vehicle Hydrocarbon Emissions, 2001*

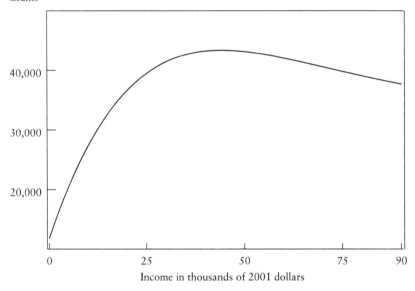

Source: Author's calculations.

per mile of driving by total predicted miles of driving yields total annual vehicle hydrocarbon emissions for each income category.

Graphing annual household vehicle hydrocarbon emissions against income produces the familiar shape of the EKC, as shown in figure 4-1. In other words, increases in household income initially lead to an increase in automotive pollution due to scale effects. However, after a turning point of roughly $40,000, further income growth causes households to generate less pollution due to offsetting quality effects.

Paying More for Green

Of course, there is no guarantee that higher incomes will translate into greener consumption. Higher-quality products are more expensive than lower quality products, and consumers will only purchase them if they believe that doing so increases their well-being. Consider the choice between conventional and organic foods. Assume that an organically grown lettuce costs $2.00, a conventionally grown lettuce costs $1.50, and the two taste the same. The environment would benefit if everyone bought the organic lettuce. Nonorganic lettuce growers would go bankrupt, sustainable farming would prosper, and fewer chemicals would go into farming and ultimately into the air and water that we share.

But how many consumers are likely to make this choice? Environmentalists are more likely to pay this green premium. People who like the idea that organic farmers are successfully competing against agribusiness might also purchase these products, as would people who value the health benefits of organic produce. But many consumers, even among the relatively wealthy, will prefer to "buy brown" and find another use for the extra 50 cents.

This phenomenon illustrates the free-rider problem. We might all be better off if sustainable farming became the norm. But many consumers are likely to reason, "I'm just one person. My consumption choices do not affect the future of sustainable farming. Given that the two lettuces taste the same, I might as well save 50 cents and purchase the nonorganic lettuce." If everyone takes this approach, the market for organic lettuce will wither, and sustainable farming will die. Recognizing this problem, environmentalists often stress that people should "think globally, act locally." This motto reflects the desire to build a culture where consumers explicitly consider the wider consequences of their actions.

How can environmentalists get around the free-rider problem? One solution is to emphasize the private benefits associated with green purchasing. In some cases these benefits are intrinsic product characteristics—the organic lettuce may actually taste better or improve your health. In other cases they are socially constructed. For example, one way to promote green consumption is by giving it social cachet. Consider the demand for hybrid vehicles, such as the Toyota Prius and the Honda Insight. These cars emit very low levels of pollution but cost significantly more than conventional cars. Consequently, they represent an excellent test of the free-rider hypothesis. If everyone bought a hybrid car, air quality throughout the United States would improve. But the private benefits associated with such a purchase are relatively small.[8] The primary private

---

8. "The hybrid delivers modestly better performance, improved mileage and slightly more space than the conventional V-6 Accord. . . . But Honda is betting that the intangible and invisible benefits of hybrid ownership will drive discriminating upper-middle income buyers to its showrooms to do their bit for the ozone layer." John M. Broder, "Greening without the Preening," *New York Times*, November 28, 2004, p. 12.

benefit is a reduction in gasoline bills, but impatient consumers are unlikely to value these future savings as highly as the $3,000 or $4,000 they could save immediately by buying a conventional car.

How many people, then, are likely to buy clean cars? The key may lie in the way they are marketed. In *The Theory of the Leisure Class*, Thorstein Veblen coined the phrase "conspicuous consumption" to refer to purchasing decisions that are driven largely by the buyer's desire to flaunt his wealth.[9] Buying a fur coat or a Rolex watch is one way for consumers to signal their status—at least to observers who care about such things. Veblen viewed such socially driven consumption as a bad thing. But this need not be the case. It all depends on which purchases signal membership in the elite.

When it comes to cars, ownership of a Mercedes or a Porsche has traditionally been a sign of sophistication—or at least of wealth. But green vehicles could potentially become status symbols as well. For example, by enlisting more Hollywood celebrities to buy and drive hybrid vehicles, environmentalists could make clean cars cool and encourage greater adoption among the public at large. Fuel-efficient vehicles could also become a means for drivers to signal their own virtue and bask in the "warm glow" effect of being seen by others as they do a good deed. As Robert Bienenfeld, a senior manager at Honda says of his company's hybrid car, "You can feel good about owning it. How do you put a price on that?"[10]

Greening Consumption among the Urban Poor

This discussion has focused on the consumption choices of the middle class and the rich, but the choices of the urban poor can also have large sustainability impacts on a growing city. As a city's economy booms, the purchasing power of the urban poor is likely to grow in response to increased demand for basic services, such as dishwashing and taxi rides. Poor households will take advantage of these income gains to alter their consumption patterns, often in ways that can improve environmental quality. For example, they may improve their diets and move to lower-density housing. In addition to delivering private benefits, both changes can improve public health.

9. Veblen (1899).
10. Broder, "Greening."

In developing countries indoor air pollution may be the most important environmental challenge that income growth among the urban poor can help solve. In India alone indoor air pollution may kill as many as one million people a year. Households create more indoor air pollution when they cook and heat their homes with dirty fuels, such as dung or wood. The very poor typically cannot afford other fuels. As a poor family's income rises, however, it has two choices for spending its growing energy budget. It can buy more of the same traditional fuels, which would increase its exposure to indoor air pollution, or it can buy higher-quality, less-polluting fuels, such as kerosene. To the extent that households choose to increase the quality, rather than the quantity, of their consumption, rising income will contribute to a reduction in pollution exposure.[11]

The Supply of Green

In order for rising income to promote greener consumption among both the urban rich and poor, there must, of course, be greener products and services for them to buy. How likely are free-market producers to supply such goods? Judging from the popular media, the answer might appear to be "not very." Hollywood, for example, routinely casts companies in an antienvironmental role. In movies such as *Silkwood, Legal Action,* and *Erin Brockovitch* and television shows such as *The Simpsons* (think of Homer's boss at the nuclear plant, Mr. Burns), profit-grubbing capitalist firms knowingly expose unsuspecting citizens to environmental risks. Rather than invest in costly pollution abatement efforts, these firms choose to conceal information about the hazards their activities create. Once an environmental tragedy does occur, they blame random events rather than negligence on the part of the firm.

But this picture is overdrawn. While firms clearly have an interest in cutting costs, they also seek opportunities to profit by exploiting new sources of demand. Many companies do extensive market research to learn what consumers want. If richer consumers want greener products, are sufficiently well informed to distinguish them from brown alternatives, and are willing to pay a green premium—admittedly, these are all big "ifs"—sellers will supply them.[12] Witness the fact that recycled toilet

---

11. Pfaff, Chauduri, and Nye (2004).
12. An open empirical question concerns how much extra people are willing to pay for a green product versus an observationally similar "brown" product. A sec-

paper competes against conventional toilet paper at supermarkets, and the Toyota Prius competes against SUVs.

It is possible to imagine a similar dynamic playing out in other areas, such as suburban development. Many environmentalists yearn for "smart growth" to take root in U.S. cities. Smart-growth communities are high-density neighborhoods where residents largely get around on foot and share public space rather than maintain their own manicured lawns. Free-market environmentalism could make this a reality. If richer people want to live in such communities, then developers have an incentive to build them.

Suburban developers have gained a brown reputation for converting pristine farmland into suburban "Levittowns." But suppose that a developer's market research reveals that potential new suburbanites would be willing to pay a premium—$400,000 instead of $300,000 for the same home—to live near a protected nature park. The prospect of this premium gives the developer a strong incentive to protect nearby environmental assets. Of course, this example hinges on the assumptions that the assets in question are beautiful and that the rich value their proximity. If the proposed development bordered a smelly wetland populated by noisy and ugly birds, free-market environmentalism would not have the same effect.

However, even in this case, market forces can still help promote environmental goals. For example, a nature trust could purchase the wetland to prevent it from being transformed into a golf course or worse. Such groups as the Land Trust Alliance, the Conservation Fund, the Nature Conservancy, and the American Farmland Trust are nonprofit clubs that collect donations from individuals and use this money to purchase land at market prices. In this way they can guarantee that land will not be converted to suburban tract housing, without raising the controversial issues of "takings" and eminent domain. In recent years announcements of open space purchases and farmer development rights purchases by such groups as the California Trust for the Public Land have become common.

---

ond open question concerns whether consumers are truly sophisticated enough to distinguish products that *claim* to be green versus those products that truly *are* green. If firms believe that consumers can be tricked, then some firms may launch green advertising campaigns without actually changing their ways of doing business. In this case sustainability will not be enhanced by rising purchase shares of "green" products.

Figure 4-2. *Employment Trends in U.S. Metropolitan Areas, 1970–2000*

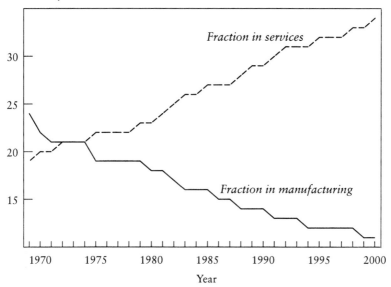

Source: Author's calculations.

Greening Urban Production

Between 1969 and 2000, the number of manufacturing jobs in New York County (Manhattan) declined from 451,330 to 146,291. Manufacturing accounted for 16.2 percent of the county's employment in 1969 compared to only 5.3 percent in 2000. Over the same period, service employment increased from 25.4 percent of the local economy to 41.1 percent. Figure 4-2 presents evidence on national trends in manufacturing and service sector employment in metropolitan areas. For 279 metropolitan areas in the United States, I calculate the share of workers employed in manufacturing and services in each year between 1970 and 2000.[13] The figure shows that manufacturing's share of employment in big cities has been monotonically falling (from 22.4 percent to 10.9 percent between 1970 and 2000) while the service sector's share has been

---

13. For each year I calculate the weighted mean of metropolitan area employment in the service and manufacturing sector using a metropolitan area's total employment as the weight.

monotonically rising. This trend is not unique to the United States. Over the last thirty years, London has lost 600,000 manufacturing jobs—and gained 600,000 jobs in business services, as well as 180,000 jobs in entertainment, leisure, hotels, and catering.[14]

Why is manufacturing leaving major cities? Increased international trade has reduced the fraction of the national workforce employed in manufacturing, and within the United States, similar factors have helped move manufacturing out of cities and into lower-cost more rural regions.[15] Rising incomes have priced many cities out of the market for manufacturing jobs. In addition, growth has made urban real estate increasingly unaffordable for industrial firms. For example, in Manhattan a 40-story skyscraper could contain 160 luxury apartments, each worth more than $2 million. No manufacturer could match a $320 million dollar offer for the same piece of land. Finally, lower transportation costs have made it feasible for manufacturers to move their operations to more distant locations and ship finished products back to urban customers.

There are some exceptions to this pattern. In particular, innovative firms in the initial stages of product development may find that urbanization economies offset the higher costs of doing business in wealthy cities. These businesses often require reliable access to very specialized workers, who are more easily found in large urban labor markets. However, once a product is fully developed and production is standardized, there is no need for many of these firms to absorb the high land and labor costs typical of large metropolitan areas.[16]

Deindustrialization's Silver Lining: The Rust Belt Turns Green

Urban deindustrialization has real costs. As stressed by the sociologist William Julius Wilson, many workers with relatively little education lose high-paying manufacturing jobs when local factories close down.[17] Many of these displaced workers end up in low-skill service positions—

---

14. For more details see Mayor of London, "The London Plan" (www.london.gov.uk/approot/mayor/strategies/sds/london_plan_download.jsp [November 2005]).
15. Crandall (1993).
16. Henderson (1991).
17. Wilson (1990).

at McDonald's, for example—that pay $10 less per hour than their previous jobs.[18] But from a green perspective, deindustrialization has important benefits. Major manufacturing industries, such as steel and chemicals, produce large quantities of toxic emissions. As these industries have moved out of the densely populated Rust Belt to new locations in the South and overseas, cities like Pittsburgh and Gary, Indiana, have experienced sharp reductions in ambient particulate levels.[19] Pittsburgh has tried to turn this quality of life gain into an engine of further growth by luring high-tech firms to work in conjunction with top-ranked local universities.

Air quality is not the only indicator of where deindustrialization has led to environmental progress. Many heavy industries were also major contributors to urban water pollution. In the late 1960s, Detroit car companies, Toledo steel mills, and Erie, Pennsylvania, paper plants all poured industrial waste into Lake Erie, turning it into a giant cesspool with 30,000 sludge worms per square foot of lake bottom.[20] In the 1920s the Inland Steel Company dumped 25 million gallons of blast furnace gas wash water and 12 million gallons of coke plant waste per day into Lake Michigan.[21] And the famous 1969 fire on the Cuyahoga River in Cleveland began when sparks from a train ignited debris that was floating in a slick of oil and chemicals. A favorite local saying at the time was, "Anyone who falls into the Cuyahoga River does not drown, he decays."[22]

When manufacturing plants close in major cities and start up in smaller southern communities, is this a zero-sum game? Is New York City's reduction in air or water pollution matched by an increase in pollution in rural Alabama? The good news is that this is highly unlikely due to technological advance. The plants that close in New York and Pittsburgh were built using older, more inefficient technologies. The new plants feature state-of-the-art technology that reduces pollution per unit of output. A prime example comes from the steel industry, where new technologies, such as minimills, produce the same output as older blast furnaces at a much lower environmental cost.

18. Neal (1995).
19. Kahn (1999).
20. "The Cities: The Price of Optimism," *Time*, August 1, 1969.
21. Baden and Coursey (2002).
22. "The Cities."

Urban Deindustrialization around the World

The migration of manufacturing away from major cities is not limited to the developed world. It has been documented globally in countries as different as Poland and South Korea. Between 1983 and 1993, there was a 35 percent decline in employment in heavy industries in South Korea's three major metropolitan areas, while the rest of the nation experienced an 85 percent increase in employment in these industries.[23] Halfway around the world, in Poland, a similar pattern has been found.[24]

The death of communism offers a dramatic illustration of how deindustrialization has delivered environmental benefits outside the United States. Under communism the state's emphasis on economic production took priority over quality of life issues such as pollution. In eastern Europe, for example, both households and industrial plants burned high-sulfur coal and fuel oil, causing particulate levels to soar.[25] These practices continued to some extent after communism's demise. In 1995 the average particulate level across nine major eastern European cities was 92 micrograms per cubic meter, whereas in twenty-one western European cities it was 59 micrograms per cubic meter.[26] However, this represented considerable improvement since 1989, when the Berlin Wall came down. Over the 1990s ambient sulfur dioxide fell by 8 percent per year in major cities in the Czech Republic, Hungary, and Poland, even when controlling for city population and national per capita income.[27] These gains provide additional support for the hypothesis that phasing out older, heavy urban manufacturing can deliver large environmental benefits since the service sector grew as a share of GDP in each of these nations during the 1990s while the manufacturing sector declined.

---

23. Henderson, Lee, and Lee (2001).
24. Deichmann and Henderson (2000). Globalization may also help push plants away from major cities. Krugman and Livas (1996) argue that in an economy that does not trade with the rest of the world's nations, a manufacturing plant is much more likely to locate in the nation's major city. The owner of this factory will reason that it is costly to open a factory and ship final output to customers. Facing these fixed costs of opening a plant and these transportation costs of shipping, the rational business person will build a large factory in the largest city. Following this strategy, the manager would only have to build one factory, and by locating it close to many of its customers in the big city, transportation costs would be economized.
25. Hughes and Lovei (1999).
26. Kahn (2003a).
27. Kahn (2003a).

These reductions in pollution are also particularly impressive because in each case per capita income remained below the conventional EKC turning point for particulates, which is roughly $7,000. For example, Poland's per capita GDP was $2,990 in 1990, $2,936 in 1993, and $4,781 in 2003.[28] (All figures are in 1995 dollars.) Yet despite this growth, ambient air pollution declined in Poland during the 1990s. One optimistic explanation for these facts is that the transition from communism to capitalism shifted the EKC for these nations down and to the left.

Conclusion: Technology, Urbanization, and Environmental Quality

Urban growth triggers offsetting effects. It increases the quantity or scale of consumption and production, but it also raises the quality of the goods consumers purchase and the techniques that producers use. The first effect reduces urban sustainability; the second often increases it. Consequently, quality upgrading as incomes rise can play a key role in greening cities.

Technological progress plays an important role in determining the possibilities for quality upgrading and the size of the resulting environmental gains. In this chapter I have chosen not to focus on technological advance, not because it is unimportant but because it is so unpredictable. In 1950 a scientist would not have been able to predict health care innovations and market products available in 2000, and the same holds true of analysts trying to peer into the future today. One famous economist once said that economists are good at predicting what happened last year.

A colorful example of how technological progress can green cities was recently reported in the *Economist*:

> Hong Kong's 10,000 restaurants produce more than 20,000 tons a year of waste cooking oil. Meanwhile its vehicles belch out so much noxious diesel soot that the air pollution index is regularly stuck at "high." Engineers there are trying to run diesel engines on this cooking oil waste. Why not dispose of one pollutant (waste

---

28. For data on Poland, see World Bank, "World Development Indicators 2006—Key Development Data and Statistics" (worldbank.org [May 2006]). This online source is the World Bank's primary database for cross-country comparable development data.

cooking oil) by burning it in diesel engines and thus reducing the level of another pollutant (vehicle fumes)?[29]

What triggers such innovation? Again, cost plays an important role. If restaurants must pay to dispose of their waste cooking oil, they will be more active in searching for cheaper or even profitable ways to dispose of it. However, it is not enough to have just the demand for new ideas and technologies; conditions must also facilitate their supply. Urban growth can help create an environment that nurtures innovation by underwriting investments in training and infrastructure. In addition, it can promote technological advances by facilitating the diffusion of new ideas.[30] This is yet another channel through which growth can foster urban environmental quality.

---

29. "Hong Kong Phooey," *Economist*, June 12-18, 2004.
30. Glaeser and others (1992).

CHAPTER 5

Income Growth and Greener Governance

Despite the forces outlined in the previous chapter, market-driven choices are unlikely to produce green cities on their own. For example, despite the tendency of richer consumers to trade up to lower-emission cars, it's hard to imagine that car manufacturers would have focused on reducing emissions to the same extent in the absence of legislation like the Clean Air Act.[1] Similarly, left to their own devices, consumers may have difficulty determining which products are truly green. Regulation, such as the Organic Foods Production Act, which sets standards for the production and processing of organic products, can help with such tasks.

For such reasons, regulation plays a key role in explaining the shape of the environmental Kuznets curve. As cities become more prosperous, both the demand for and the supply of environmental regulation grow. As noted previously, economic development typically increases access to education, which tends to foster greener preferences. It also changes urban production patterns in a way that promotes greater investments in quality of life. Finally, income growth gives politicians both the incentives and the means to make urban sustainability a significant policy priority.

The results at the national level are reflected in such measures as the Environmental Sustainability Index (ESI), which was designed by

---

1. However, some progress might have been made even in the absence of such legislation. Kahn and Schwartz (2006) document that vehicle emissions fell in model years when emissions regulation was not tightened.

researchers at Columbia and Yale.[2] The 2005 ESI ranks Finland, Norway, Uruguay, Sweden, Iceland, Canada, and Switzerland at the top with respect to its overall sustainability criteria among 146 nations, while relegating Haiti, Uzbekistan, Iraq, Turkmenistan, Taiwan, and North Korea to the bottom of the list. The United States was ranked forty-fifth. This relatively low score was mainly due to the high level of U.S. greenhouse gas emissions. One component of the ESI ranking is environmental governance. Based on this criterion, the top five nations are Finland, Sweden, Norway, Switzerland, and Japan. The United States ranks fourteenth. The five nations with the lowest environmental governance scores are Angola, Iraq, Liberia, Uzbekistan, and Turkmenistan. This suggests that broadly speaking, richer nations outperform poorer ones when it comes to designing and implementing green policies. Similarly, other cross-national studies have reported that wealthier nations have higher-quality regulatory institutions.[3]

Why Is Government Intervention Necessary?

Before investigating the forces behind greener governance, it is worth asking why government intervention is necessary at all. As individual preferences become greener in response to income growth, why can't the market supply what consumers increasingly demand? One way to think about this problem is to consider what a city without government would have to look like in order to remain green. Imagine a city with only two residents and let each resident own half the city's land. Assume that litter is the city's only environmental problem and that detection of litter is costless. Under these conditions, if landowner A's dog drops its dirty bones on landowner B's property, B will immediately be aware of the problem, and A will have to compensate B. No government intervention is necessary. Instead, the two landowners can bargain their way to a mutually beneficial agreement.[4]

---

2. The ESI is the result of collaboration among the World Economic Forum's Global Leaders for Tomorrow Environment Task Force, the Yale Center for Environmental Law and Policy, and the Columbia University Center for International Earth Science Information Network (CIESIN). See Yale Center for Environmental Law and Policy, "Environmental Performance Measurement Project" (ww.yale.edu/esi [May 2006]).
3. La Porta and others (1999); Dasgupta and others (2004).
4. Anderson and Leal (2001) provide many compelling examples where bargaining solves the environmental externality problem. It is important to note that almost

This example relies on several key assumptions. First, there are no asymmetries of information. Landowners A and B are equally well informed of any pollution that occurs. Second, there are no transaction costs. Identifying and negotiating with polluters is free. Third, there are no externalities. If landowner A lets litter accumulate on his own property, none of this pollution drifts over the property line and imposes itself on landowner B. (This presupposes, of course, that the sight and the smell of the litter are not concerns.) Finally, property rights are well defined. There is no dispute about landowner B's right to keep his property free of dirty bones.

In practice, these conditions are rarely met. It is not always possible to identify a polluter and hold that person or entity accountable, and doing so is almost never free. In fact, people may not even be aware that pollution is occurring at all or understand the impact that it can have. In addition, negative externalities are a pervasive problem where pollution is concerned. The owner of a factory that pours smoke into the air is not the only one to suffer from the resulting smog. The rest of the city also suffers, which means that, in the absence of regulation, the plant will pollute far past the point where the marginal social benefit equals the marginal social cost.

Finally, in many cases private property rights do not exist. This is true, for example, of city air, streets, subways, and often parks. Since these are common property, no one can stop another resident from, say, dropping a used cigar on the ground. This problem is known as the tragedy of the commons.[5] No one likes litter, but no one has a sufficiently strong incentive to invest in combating it on public property. Worse, in the absence of effective regulation, everyone has an incentive to keep littering even though they wish that everyone else would stop.

In isolated cases private actors may succeed in banding together to solve such problems. Sometimes a single individual may even find it in

---

all of the cases they refer to are based on natural resource allocation in less populated rural areas. In such settings it is relatively easy for the victim of pollution to identify the perpetrator. This facilitates Coasian bargaining. (For more on the economic theory of Ronald Coase, see nobelprize.org/economics/laureates/1991/press.html [May 2006].) In cities, by contrast, high transaction costs can make such bargaining infeasible. Imagine, for example, that Bill Gates were asthmatic and willing to pay Seattle polluters not to pollute. He would still face the challenge of finding and bargaining with each polluter.

5. See Hardin (1968).

his interest to step into the breach. For example, if Donald Trump believed that cleaning up Central Park would raise the value of his property, it would make sense for him to contribute to keeping the park clean. If the increase in his property's value were sufficiently large, it would make sense for him to bear the entire cost. Along similar lines, in some downtown areas, local businesses have formed business improvement districts. These groups collect resources to be spent on garbage cleanup and general investments to make the neighborhood more attractive for business activity. However, the free-rider problem makes it difficult to apply such solutions on a broader scale.

The Demand for Green Policies

In an ideal world, benevolent politicians might track issues, such as air pollution, and adopt appropriate regulation once the benefits of intervention exceed its costs. Actually, conducting such a cost-benefit analysis is part art and part science. Researchers must estimate how much a particular regulation would cause environmental quality to improve, as well as the value that people would put on this improvement.[6] They also have to measure the costs of regulation, which is an extremely complicated task.

But the larger problem is the fact that politicians are not always benevolent, and the technical costs and benefits of various policies are rarely their primary concern. Instead, many politicians are preoccupied with getting reelected. This preoccupation generally does not favor environmental legislation. Free-rider problems reduce the likelihood that pro-environment candidates will collect large donations from households, but electric utilities, chemical companies, and other firms have a

---

6. The public health methods and hedonic valuation methods discussed in chapter 2 could be used to value the marginal improvements in environmental quality. Measuring how much pollution has declined as a result of regulation would require constructing a "counterfactual." Suppose that a city that has increased its regulation experiences rising pollution levels. It could be the case that this regulation has been effective at reducing pollution. The key unknown is what would pollution levels have been had the regulation not been enacted. For example, if a city experiences a five-unit increase in pollution during a time period when it enacted regulation, but it would have experienced a fifteen-unit increase in pollution had it not enacted this regulation, then regulation caused a ten-unit reduction in pollution. For a study of regulation's effects on ambient ozone levels in the United States, see Henderson (1996).

strong interest in funding candidates who will oppose regulation that could reduce their profits. Using data from the Federal Election Commission, I have shown that all else equal, candidates that receive a larger share of their campaign contributions from private companies are more likely to vote against environmental regulation.[7]

How can income growth counter this effect? As the following section shows, environmental regulation is more likely to rank high among urbanites' demands as per capita incomes rise. Wealthier voters are also more likely to monitor elected officials closely to see how they are governing on issues such as crime and the environment. This gives politicians less scope to pursue their own narrowly self-interested agendas.

Education and the Demand for Green Governance

By boosting educational attainment, urban economic growth helps put environmental issues on the political agenda in several ways. People with more education tend to be more interested in quality of life issues, such as the environment, that go beyond bread-and-butter concerns.[8] People with more education tend to be more patient and more likely to support costly investments that address long-run environmental threats.[9] More educated people are more likely to demand in-depth analysis of environmental issues. This sets a virtuous cycle in motion by providing incentives for the media to research and present stories on pollution and the environment. Finally, people with more education tend to play a more active political role. For example, educational attainment is positively associated with the likelihood that a person votes.[10]

Data from voting on binding ballot initiatives in California provide an opportunity to test the link between education and green policy preferences. Such initiatives are translated directly into legislation if they

---

7. Kahn (2006).
8. Of course, all environmental problems are not created equal. Public health concerns, for example, are a much more salient issue with urbanites than the more abstract notion of a growing ecological footprint. Tell a city's residents that the city's footprint has grown 40 percent over the last decade, and there is unlikely to be a rush to the mayor's office demanding action. Contrast this with the outcry that would occur in the event of a public health crisis, such as an outbreak of severe acute respiratory syndrome (SARS).
9. Becker and Mulligan (1997).
10. Moretti (2004).

receive a majority vote. Over the last thirty years, voters in California have had the opportunity to vote on a wide array of environmental issues, including increasing expenditure on public transit, raising gasoline taxes, weakening antismoking laws, selling bonds to improve the quality of the water supply, and improving local air quality. By matching voting data to data on local educational achievement (at both the census tract and the county levels), I have shown that areas with a higher proportion of college graduates are consistently more likely to support the pro-environment position.[11]

Similar evidence comes from congressional voting records. Since 1970 the League of Conservation Voters (LCV) has published an annual scorecard that rates each legislator's environmental record.[12] These ratings are based on votes that a panel of experts has identified as the most important environmental issues of the year. For example, in 1998 the LCV focused on thirteen environmental votes on such issues as takings, logging in national forests, Alaskan logging roads, Gulf of Mexico fisheries, restricting environmental protections, energy efficiency programs, the global warming gag rule, and tropical forest conservation. A legislator voting the pro-environment position on six of twelve bills would receive a score of 50 percent. Using data from 1970 to 1994 on the average LCV score for each state's congressional delegation, I have found that legislators from states with higher educational levels are more likely to vote pro-environment. A 10 point increase in the percentage of adults who are college educated increases pro-environment voting by 11 percentage points.[13]

Other Forces Driving Green Policy Demand

Urban economic growth can also increase demand for greener policies even if average educational levels do not change. For example, as incomes rise, more people are likely to own their homes. This development creates a stakeholder class with a strong incentive in preserving neighborhood environmental quality. Unlike renters, who often migrate away and have no financial stake in the community, homeowners realize

---

11. Kahn and Matsusaka (1997); Kahn (2002).
12. See the League of Conservative Voters, "National Environmental Scorecard" (www.lcv.org [April 2006]).
13. Kahn (2002).

that their property will lose value if local environmental quality declines.[14]

Sometimes, of course, this realization can have unfortunate effects. Every city faces the question of where necessary evils, such as garbage dumps, should be located. Richer communities with more homeowners are more likely to band together to lobby politicians not to situate such noxious facilities in their neighborhoods.[15] Anticipating such a response, forward-looking politicians often place such facilities in poorer renter communities. In this case homeownership greens the local community, but it is a zero-sum game.[16]

Individuals are not the only urban actors whose interest in the environment grows as incomes rise. As urban economies deindustrialize, they are increasingly dominated by service sector firms with a strong stake in local quality of life. Highly skilled, creative workers tend to be footloose. While a lumberjack must live near the forest that he cuts down, a computer programmer can work just about anywhere with a computer and a high-speed Internet connection, and can write, debug, and ship code to customers far away. Consequently, firms must compete for such workers by offering not only higher salaries but also a higher quality of life.[17] This makes them natural advocates for local environmental regulation.

---

14. Such a scenario played out recently when the entertainment mogul David Geffen sought to keep the public from having access to the Pacific Ocean near his Malibu house. If he had won his case, then environmentalists would have had mixed feelings. As stakeholders Geffen and his fellow wealthy Malibu beachfront friends would be a powerful interest group in preserving this natural capital, but the general public would have less beach access. A philosophical issue arises: how much good is achieved by protecting natural capital if very few people have access to it? The California Coastal Commission issued a cease and desist order against the property owners, directing them to end the posting of no trespassing and private property signs. The public now has beach access through paths near these expensive homes. See Surfrider Foundation, "Beach Access" (www.surfrider.org/malibu/projects.htm #access [April 2006]).

15. Olson (1965).

16. An environmental justice advocate would point out that the community that receives the dump has not been explicitly compensated for receiving the dump. This boils down to a property rights issue. If each community had a veto right such that it could veto the placement of a dump in its vicinity, then one community would receive the dump plus compensation from the other communities that chose to pay to reduce their exposure to waste.

17. Florida (2002).

Cities that attract high-skilled, creative workers typically experience greater economic growth.[18] Consequently, urban politicians are increasingly aware of the need to provide the lifestyle advantages that such workers demand.[19] Mayors who do not care about the environment—but do care about their tax base—will become environmentalists if they sense that skilled workers value such amenities and are likely to "vote with their feet." This is particularly important in richer nations, which typically delegate more power to local decisionmakers and consequently give them more opportunities to compete for residents.[20]

In addition, in many postindustrial cities, the tourism sector is a growing employer. This industry represents another powerful force lobbying to preserve quality of life in urban areas. Growing interest in tourism, for example, has inspired cities ranging from Boston to Providence to Nashville to rediscover their waterways as urban amenities. As a result of this shift in perspective, many former dumping grounds have become valuable resources that local actors seek to rehabilitate and preserve.

Supplying Greener Governance

By using their powers of taxation, regulation, and zoning, government authorities can play a key role in greening a city. Whether a government is up to this job of protecting the environment depends on the incentives facing elected officials and on the resources they control. Richer cities are more likely to have access to the expertise necessary to design effective policies. They can also invest in the monitoring and enforcement that are necessary, along with a well-functioning judicial system, to make regulation work.[21] Below I examine four areas in which economic

---

18. Glaeser and others (1992).
19. Florida (2002); Glaeser, Kolko, and Saiz (2001).
20. Henderson (2002) argues that richer nations decentralize more political power.
21. A well-functioning judicial system signals polluters that they will be held accountable for damage caused to the urban environment. In the United States, the torts system has made many polluters pay millions of dollars for past environmental damage, which makes potential polluters think twice before committing the same offense. If a firm or other entity anticipates a large fine for polluting, then it is more likely to take precautions to minimize the likelihood of a costly event, such as the *Exxon-Valdez* oil spill. But in countries where courts are known to be corrupt, firms have little incentive to green their production. They can simply conduct business as usual and pay off a judge if they get caught.

development has contributed to improvements in urban environmental quality in the United States. In some cases the relevant regulation has been passed at the national level; however, enforcement has been concentrated in major cities.

Air Quality

Government regulation can affect a city's air quality by influencing both the scale of local economic activity and the technologies used. There are various ways to achieve these goals. For example, raising gasoline taxes, which are much lower in the United States than in Europe, would reduce the scale of gasoline consumption in the short term.[22] In the medium term, it would also encourage vehicle producers to invest in fuel-efficient technologies. To date, however, U.S. policymakers have chosen to rely on direct mandates or "command and control." The centerpiece of this approach is the 1970 Clean Air Act, which focuses on reducing the emissions generated by new additions to the capital stock. For example, due to Clean Air Act mandates, newer coal-fired electric utility boilers produce considerably fewer pounds of nitrogen oxide per megawatt-hour than older models. The average emissions rate of units built before the New Source Performance Standards were adopted in the 1970s was about six pounds of nitrogen oxide per megawatt-hour. This figure dropped to 4.09 in the 1980s and 3.55 in the 1990s.[23] These improvements are important because the electricity sector accounts for roughly 20 percent of the nitrogen oxide emitted in the United States.

Another example of regulation-induced technological change can be seen with vehicle emissions. Vehicles built after 1975 were required to produce 72 percent less hydrocarbon emissions than older models. This

---

22. An ongoing research question is determining what would be a socially optimal gasoline tax. One study concludes that the U.S. tax is too low and the European tax is too high (Parry and Small 2005). Do persistent low gas taxes indicate that U.S. citizens do not care about consequent environmental problems associated with gasoline consumption? Pietro Nivola and Robert Crandall argue that this inference is not valid. Instead, differences between gasoline tax rates in the United States and Europe largely reflect differences in how the revenue generated by such taxes is spent. In the United States, all gasoline tax revenue is reserved for particular public works, whereas in Britain and France, all taxes are treated as general revenue. See Nivola and Crandall (1995, p. 69).

23. Burtraw and Evans (2003).

Figure 5-1. *Distribution of Ambient Ozone by Monitoring Station, California, 1980–2001*[a]

Parts per million

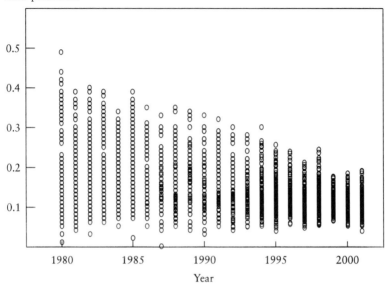

Source: California Air Resources Board (2004).
a. Maximum one-hour reading.

measure has had a dramatic impact on air quality, as data from California show. Figures 5-1 and 5-2 present data from 1980 to 2001 on maximum readings by monitoring station by year for the two pollutants that are mainly produced by vehicle emissions: ozone and oxides of nitrogen. Time trends in the distribution of these statistics show steep progress. For both pollutants the entire distribution is shifting downward and compressing over time. These trends are particularly significant because of the role that cars play in contributing to air pollution. According to California's 1995 Statewide Emissions Inventory Summary, mobile sources are responsible for 28 percent of total organic gas emissions, 82 percent of carbon monoxide emissions, 80 percent of nitrogen dioxide emissions, and 48 percent of sulfur dioxide emissions.[24]

Overall, there has been remarkable improvement in U.S. air quality over the past thirty-five years. In 1970 the average total suspended par-

24. See California Air Resources Board, "1995 Statewide Emission Inventory Summary" (www.arb.ca.gov/aqd/almanac/almanac99/pdf/tbl2_01.pdf [May 2006]).

Figure 5-2. *Distribution of Ambient Nitrogen Dioxide by Monitoring Station, California, 1980–2001*[a]

Parts per million

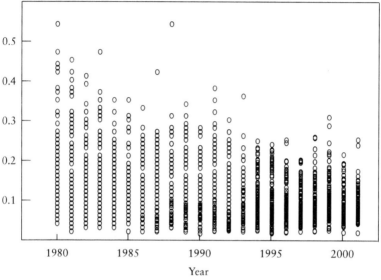

Source: See figure 5-1.
a. Maximum one-hour reading.

ticulate level in the United States was 93.3 micrograms per cubic meter. By 1978 the average level had fallen to 69.4 micrograms per cubic meter.[25] This progress continued in later years. Nationally, between 1988 and 2003, particulate matter 10 microns and less in diameter (known as PM-10) fell by 31 percent.[26] Today, the average U.S. city has much better air quality, as measured by levels of ambient particulates, nitrogen oxide, sulfur dioxide, ozone, and carbon monoxide, than thirty years ago. For example, ambient ozone levels have declined by 26 percent between 1980 and 2004.[27] However, in 2003 there were still 209 counties containing 100 million people where ambient ozone levels exceeded the eight-hour National Ambient Air Quality Standard under the Clean Air Act.[28]

Despite these gains, the command-and-control approach embodied in Clean Air Act regulation has some drawbacks. For example, requiring

25. Chay and Greenstone (2003).
26. See U.S. Environmental Protection Agency (EPA, 2004).
27. EPA, "Air Trends—Ozone" (www.epa.gov/airtrends/ozone.html [April 2006]).
28. EPA, "Ozone Report—A Look at 2003" (www.epa.gov/airtrends/2003ozonereport/ lookat2003 [April 2006]).

polluters, such as electric utilities, to install specific emissions-reducing devices does not allow remedies to be tailored to different situations. Economists have pushed regulators to consider alternative regulatory schemes that could provide stronger incentives for urban polluters to invest more money in research and development on greener production techniques. One such alternative is a pollution permit system, which would require firms to pay for the right to pollute. The price of this right would be determined by supply and demand, with policymakers determining the aggregate supply. For example, if scientists believe that a city's atmosphere can absorb 500 tons of emissions without causing a significant increase in morbidity and mortality risk, the city could auction off 500 tons of pollution permits to the highest bidders.

The U.S. sulfur dioxide trading permit market demonstrates the benefits of such a flexible regulatory program.[29] Unlike a command-and-control system, permit trading can be designed to reward firms that reduce their own emissions. For example, if each firm is given the right to emit five tons of soot, a company that emits only two tons can sell another firm the right to emit the other three tons. This creates an incentive for profit-maximizing firms to "green" their production process.

Properly enforcing a pollution permit system requires significant government expenditure, which is one reason such policies become more likely as incomes increase. The regulator must have the resources to engage in spot audits of polluters, to detect which industrial plants are "super emitters," and to credibly commit to fine such companies if they are caught out of compliance. When the U.S. sulfur dioxide market was created in the early 1990s, electric utilities knew that their emissions would be measured and recorded every second of the day and that regulators would impose high fines on utilities caught polluting without a permit. This greatly reduced the incentives to cheat. In contrast, a regulatory authority that is unable to precommit to impose high expected costs on urban polluters will rightly be seen as a paper tiger and its actions will have little or no effect.

UNINTENDED CONSEQUENCES OF AIR QUALITY REGULATION. While the Clean Air Act has had a positive effect on air quality, it has also had important unintended consequences. By focusing regulatory efforts on newer capital, it has aged the capital stock. By keeping an older car or running an older, less-regulated plant, a decisionmaker can

29. Stavins (1998).

delay paying the "regulatory tax" that is imposed on new capital.[30] Similarly, by focusing regulatory efforts on more polluted areas, these regulations have encouraged the migration of dirty activity to areas that are relatively pristine.[31] The Clean Air Act is not uniformly enforced across counties. Counties where pollution exceeds Clean Air Act pollution standards are called "nonattainment" counties, and clean counties are called "attainment" counties. Attainment counties, like old capital, face less regulation. As a result high-pollution plants are relocating to these areas.[32] Is this a bad trend? It undoubtedly seems that way to rural residents facing industrial pollution and waste for the first time. But from a public health perspective, the diversion of manufacturing away from highly populated cities toward the countryside is likely to reduce the total sickness produced by any one factory. A city of one million people has many more potential victims from a steel plant's soot than a rural area with 2,000 residents.

Efforts to meet Clean Air Act standards have also produced more pollution in some cases. For example, oil companies began to add methyl tertiary-butyl ether (MTBE) to gasoline in the late 1970s. MBTE helps gas burn more cleanly. However, this chemical can also contaminate groundwater when pipelines, fuel tanks, and other containers or equipment leak, and when fuel is spilled. MTBE has been widely detected in groundwater and surface water in southern California. In Santa Monica contamination levels have limited the use of most local wells. This has increased the city's dependence on imported water and treatment systems. In response to this problem, California recently phased out the use of MTBE.[33]

Finally, in some communities reductions in local air pollution have led to gentrification. As smog in, say, eastern Los Angeles declines, more people seek to move to these neighborhoods, driving up rents. Homeowners in these communities enjoy an appreciation of their housing property, but the poorest renters may be forced to move out. Only in

30. Gruenspecht (1982).
31. This grandfathering creates an odd situation: incumbent firms actually have an incentive to lobby for more regulation since this acts as a barrier to entry for new firms and increases the monopoly power of incumbents.
32. Becker and Henderson (2000); Greenstone (2002); Henderson (1996); Kahn (1997).
33. Department of Water Resources, "Chapter 5: South Coast Hydrologic Region," *California Water Plan Update 2005. Volume 3: Regional Reports* (www.waterplan.water.ca.gov/docs/cwpu2005/vol3/v3ch05.pdf [April 2006]).

rare cases will neighborhoods experience a "free lunch" of improvements in local environmental quality without an increase in housing prices.[34]

REGULATION AND ENVIRONMENTAL JUSTICE. Related to the issue of gentrification is the broader question of who benefits from environmental regulation. In much of California, there has been a tremendous reduction in air pollution over the last thirty years. Who have been the big winners from this improvement? Have the rich reaped most of the benefits of regulation, or have they been shared in equal measure by the poor?

To answer these questions, one needs to look at where the rich and poor are likely to live. Throughout the United States, the population often segregates by race and income.[35] Rich whites tend to live in the suburbs of major cities whereas poor blacks and Hispanics live in the center cities.[36] Consequently, even within the same metropolitan area, the average white resident may be exposed to a very different level of pollution than the average black or Hispanic urbanite. Effective regulation can reduce these disparities by improving environmental quality in the communities with the worst pollution problems, where the poor and minorities tend to be overrepresented. Consider the fact that effective smog regulation can only push the number of high-smog days down to zero. As a result areas that always had low smog levels will benefit less from such regulation than neighborhoods where smog levels have historically been high.

To investigate how Clean Air Act regulation has affected the pollution exposure of different demographic groups, I use evidence from California, which has an extensive ambient air pollution monitoring system. A CD-ROM data set distributed by the California Air Resources Board provides information on levels of five of the six ambient air pollutants regulated under the Clean Air Act—carbon monoxide, nitrogen dioxide, ozone, particulate matter, and sulfur dioxide—at monitoring stations located in high-population areas around the state.[37] I combine

---

34. For example, if information about the improvements in local air quality only slowly reaches outsiders or if it is costly to migrate to these areas, then incumbent renters may not experience rising rents as local quality of life improves.

35. Cutler, Glaeser, and Vigdor (1999).

36. Glaeser, Kahn, and Rappaport (2000).

37. See California Air Resources Board (2004). This CD-ROM provides all air quality readings taken in the state from 1980 to 2002. In this data set, the unit of analysis is a monitoring station.

this information with census data to determine average pollution exposure by demographic group.[38] Assume that there are only two census tracts, A and B. In this case I would use the following method to calculate pollution exposure for a particular group:

$$\text{Exposure} = (\text{share who live in A})*(\text{pollution in A}) \\ + (\text{share who live in B})*(\text{pollution in B}).$$

For example, if 40 percent of people over age sixty-five live in an area where pollution is 100 units, and 60 percent of people over age sixty-five live in an area where pollution is 0 units, then the average elderly resident is exposed to 40 units of pollution. The calculations presented below are based on a generalization of this equation reflecting the fact that there are thousands of census tracts. In these calculations the population shares sum to one.

Table 5-1 shows how population pollution exposure in California changed between 1980 and 1998 (assuming that the population was distributed in both years as it was in 1990). Reading across a row reveals differences in pollution exposure for different demographic groups; reading down a column reveals differences in exposure for the same demographic group across different pollutants and years. In both 1980 and 1998, wealthier households were exposed to less pollution than poorer households. But the difference diminished significantly over time. In 1980 the average person who lived in a census tract where the median income was greater than $65,000 (in 1990 dollars) was exposed to 25 percent less nitrogen oxide than people who lived in tracts where the median income was less than $30,000. The wealthier households were also exposed to 33 percent less carbon monoxide and enjoyed six

---

38. Summary Tape File 3A from the 1990 Census of Population and Housing contains demographic information (median income, percentage of residents who are college graduates, and the like) for census tracts, which are small geographic units including roughly 4,000 people. This data set also provides the latitude and longitude of each census tract. See U.S. Census Bureau, "1990 Census" (www.census.gov/main/www/cen1990.html [May 2006]). The California air pollution data provide the latitude and longitude of each monitoring station (California Air Resources Board 2004). The two data sets are merged by using geographic information software and census tract information to calculate the distance between each monitoring station and each census tract. If a monitoring station lies within 8,000 feet (less than two miles) of a census tract, then this tract is included in the data set. Otherwise, it is dropped from the analysis. This procedure allows creation of a data set with roughly 1,800 census tracts including over 7 million California residents.

Table 5-1. *Trends in Exposure to Ambient Air Pollution by Demographic Group, California, 1980 versus 1998*[a]

Parts per million unless otherwise indicated

Pollutant	Year	Income level in 1990 dollars			Ethnicity		
		All	<$30,000	>$65,000	White	Black	Hispanic
Carbon monoxide	1980	8.775	10.792	7.134	7.89	11.223	10.658
	1998	4.132	5.312	3.531	3.772	5.039	5.059
Nitrogen dioxide	1980	0.169	0.191	0.144	0.160	0.174	0.195
	1998	0.085	0.097	0.078	0.081	0.090	0.096
Ozone	1980	0.103	0.098	0.101	0.106	0.081	0.111
	1998	0.07	0.067	0.071	0.072	0.060	0.069
Number of high-	1980	31.157	31.015	25.213	31.799	16.764	40.416
ozone days	1998	4.479	3.935	4.587	4.770	3.197	4.685
Sulfur dioxide	1980	0.006	0.008	0.005	0.006	0.008	0.007
	1998	0.002	0.002	0.002	0.002	0.002	0.002
Particulate matter (micrograms per cubic meter)	1998	49.266	57.791	44.516	48.593	47.602	54.142

Source: Kahn (2001b).

a. Carbon monoxide values are the average of the top thirty maximum eight-hour concentration measurements. Nitrogen dioxide values are the top daily maximum one-hour concentration measurements. Ozone is measured as the average of the top thirty daily maximum eight-hour concentration measurements. "High-ozone days" is the count of days exceeding the Clean Air Act's national one-hour standard. Sulfur dioxide values are the average annual arithmetic mean. Particulate matter values are the average of the ten highest daily measurements during the year.

more low-ozone days. By 1998 all three gaps had shrunk. For example, in 1980 poor communities were exposed to 0.047 parts per million more of nitrogen dioxide than richer communities, but by 1998 the difference had fallen to 0.019 parts per million.

Table 5-1 also shows trends in pollution exposure by ethnic group. Between 1980 and 1998, Hispanics enjoyed the greatest reductions in pollution exposure. For example, exposure to high ozone days fell by thirty-six days for Hispanics, fourteen days for blacks, and twenty-seven days for whites. In 1998 Hispanics and whites were both exposed to roughly five high-ozone days, compared to forty for Hispanics and thirty-two for whites in 1980. These findings suggest that Clean Air Act regulation has helped increase environmental justice.[39]

Of course, a pessimist could rightly point out that even today urban households are exposed to too many days of bad air. This raises a fascinating economic question: now that average high-ozone exposure has

39. Kahn (2001b).

fallen below five days per year, how much would it cost at the margin to lower exposure by an extra day, and how much would the average household value this extra day of clean air? Economic theory predicts that the marginal cost of reducing air pollution rises as regulators try to achieve more ambitious goals. Moreover, as air quality improves, the marginal benefit of having an extra day of clean air is likely to decline.[40] Given existing driving, industrial, and energy production technologies, few people in California would vote for reducing smog to zero. Achieving this ideal would require shutting down the economy.[41]

Water Quality

Water quality is a second area where regulation has clearly paid off. Since 1972 enforcement of the Clean Water Act has increased the number of swimmable miles of rivers and streams in populated areas by 12,527 miles, fishable miles by 16,727 miles, and boatable miles by 14,653 miles.[42] This progress has been achieved by implementing water pollution control programs targeting sewage treatment plants and industrial sources, such as chemical manufacturers, pulp and paper mills, steel plants, and food processing plants. In addition, regulation has tackled water pollution from runoff from impervious surfaces, landfills, and hazardous waste sites.

Even more important gains were achieved in the century preceding enactment of the Clean Water Act thanks to government efforts to reduce the dangers posed by water-related diseases. In the 1880s the average urbanite in the United States lived ten fewer years than the average rural resident.[43] Between 1880 to 1940, investments in water filtration and purification played a significant role in eliminating this "urban death premium," as Louis Cain and Elyce Rotella report: "Government spending lowered U.S. urban death rates from typhoid, dysentery and diarrhea

---

40. In major cities such as Los Angeles, communities differ with respect to their air pollution levels. As discussed in chapter 2, households that value living in a very clean community, perhaps because a household member has asthma, can pay a housing premium to move to such a community.
41. The Clean Air Act does not allow the Environmental Protection Agency to explicitly engage in cost-benefit analysis. The EPA is charged with achieving the National Ambient Air Quality Standards for six measures of ambient pollution, regardless of the cost.
42. EPA (2000).
43. Haines (2001).

from 1900–1930 as cities increased their expenditures on water supply, sewage disposal and refuse collection. Between 1902 and 1929, death from waterborne diseases fell by 88 percent. In the year 1902, waterborne diseases accounted for 8.9 percent of all urban deaths while by 1929 its share had declined to 1.4 percent."[44] Based on data on mortality rates and sanitation expenditures for forty-eight U.S. cities with more than 100,000 residents, Cain and Rotella conclude that "a one percent increase in expenditure would have saved 18 lives annually in the average sized city. Cities could and did buy themselves lower death rates."[45]

Further gains in improving urban water quality often require extensive resources. For example, richer cities can protect their water supply by creating "moats" of undeveloped land around key reservoirs, as New York City has tried to do by spending $260 million to acquire additional land upstate.[46] The city has also invested $232 million since the early 1990s in upgrading the eight upstate sewage treatment plants that it owns and operates in the 1,969-square-mile watershed that extends 125 miles to its north and west. In addition, it has spent around $240 million to rehabilitate and upgrade city-owned dams and water supply facilities in the area.

Similarly, the city governments of Santa Monica and Los Angeles have jointly established the Santa Monica Urban Runoff Recycling Facility (SMURRF), a state-of-the-art facility that treats dry weather runoff water before it reaches Santa Monica Bay. Urban runoff, which is generated by improper disposal of waste and leaky septic systems, among other sources, is a growing problem and one that can be particularly difficult to combat since identifying individual culprits is an enormous task.[47] SMURRF can handle up to 500,000 gallons of urban runoff per day and has been funded in part by Santa Monica's storm water utility fee, which generates about $1.2 million a year. Instituted in 1995, this fee helps finance a variety of programs to reduce or treat

44. Cain and Rotella (2001, p. 139).
45. Cain and Rotella (2001, p. 147).
46. Archives of the Mayor's Press Office, "Mayor Giuliani and Governor Pataki Announce Final Watershed Accord: Landmark Three-Point Package to Protect City's Drinking Water (September 10, 1996)" (www.nyc.gov/html/om/html/96/sp431-96.html [April 2006]).
47. Department of Water Resources, "Chapter 21: Urban Runoff Management," *California Water Plan Update 2005. Volume2: Resource Management Strategies* (www.waterplan.water.ca.gov/docs/cwpu2005/vol2/v2ch21.pdf [April 2006]).

runoff, including street sweeping, trash collection, sidewalk cleaning, and equipment purchases and maintenance.

In addition to investing in better infrastructure, richer cities can also create incentives for conservation by making customers pay for each gallon of water they consume.[48] To institute this type of pricing, cities must install water meters and fund systems for billing and a wide range of related tasks. All of these tasks require resources that a cash-strapped government may not have. But they can have a dramatic impact on water use. In Denver, for example, environmentalists were able to block production of the Two Forks dam by demonstrating that the use of meters and water-saving devices could save more water than Two Forks would provide—and cost about 80 percent less than the $1 billion dam. Following the decision by the Environmental Protection Agency (EPA) to veto the Two Forks project in 1990, city investments in conservation paid off. Average household water use declined by 9 percent in two years.[49]

This example highlights the role that expectations of future resource prices play in determining a city's ecological footprint. Denver politicians, firms, and residents who expected a federally subsidized dam to lower the price of water had little incentive to reduce their consumption. Once the dam was no longer an option, investing in green policies made more sense. Unfortunately, this second outcome may become rarer as urban growth translates into increased political clout. A growing area with more voters and thus a larger congressional delegation is more likely to be successful at obtaining federal funding for major projects. Such political transfers ease pressures to reduce or stabilize a city's resource use.

Solid Waste

Richer cities also suffer less pollution from solid waste, even though they produce more garbage per capita than poorer cities, because they

---

48. Cities would be greener if residents had to pay per gallon of water consumed, bag of garbage created, or rush hour trip taken downtown. As an extension of this idea, economists have advocated "peak toll" pricing. Under this scheme the price of water would rise during droughts, and highway tolls would increase during rush hour. The public has typically rejected this product differentiation, even though people are used to spending different prices for the same good in other settings. For example, air travel is more expensive if you do not stay over a Saturday night.

49. McCully (1996, p. 10).

have the resources to dispose of waste effectively.[50] As a consequence, from the public health and hedonic perspectives, one would say that a city "solves" its garbage problem as it grows richer; however, an ecological footprint analysis would find that its sustainability declines.

Perhaps the simplest way of disposing of garbage is to sell it to another location. This strategy breaks the link between consumption and exposure to the resulting waste. As major cities become richer, they often become net exporters of garbage. As more cities take this route, increasing demand for trash disposal causes prices to rise. At the same time, growing environmentalism often limits the supply of available landfills and encourages communities near landfills to demand greater compensation for taking others' trash. These trends all create incentives to generate less waste.

Consider the example of New York City. In the 1960s more than 17,000 apartment building incinerators and 22 municipal incinerators burned one-third of the city's trash. The remaining garbage went to Fresh Kills and other landfills in the outer boroughs.[51] Over time, public pressure for alternative disposal options began to mount. Old incinerators and landfills were gradually shut down. The last municipal incinerator closed in 1992, and by the late 1990s, Fresh Kills was the only active dump. At the time, New York City was sending 13,000 tons of garbage per day to Fresh Kills at a price of $50 a ton. After Fresh Kills closed in March 2001, the city began to ship waste to Virginia at the cost of $120 a ton. This 140 percent price increase created a powerful incentive for New York City to economize on its production of garbage. Between 2000 and 2005, there has been a slight reduction in tons of waste disposed of per day despite the city's economic growth.[52]

One way to shrink a city's production of garbage is to administer and enforce sophisticated pricing schemes. There are many different ways to charge urbanites for services such as garbage collection. The simplest

50. Beede and Bloom (1995).
51. See John McCrory, "The First Regional Government Still Cries for Planning: The Case of Waste Management," *Progressive Planning*, March-April 1998 (www.plannersnetwork.org/publications/1998_128/McCrory.htm). See also Earth Engineering Center and Urban Habitat Project, "Life after Fresh Kills: Moving beyond New York City's Current Waste Management Plan" (www.seas.columbia.edu/earth/EEC-SIPA-report-NYC-Dec11.pdf [December 2001]).
52. See Marc V. Shaw, Peter Madonia, and Susan L. Kupferman, "Department of Sanitation," *Mayor's Management Report* (www.nyc.gov/html/dsny/downloads/pdf/guides/mmr/dsny0904_mmr.pdf [September 2004]).

method is to collect taxes on sales, income, and property and fund services out of general revenue. The problem with this approach is that it provides no incentive for households to economize on garbage production. Alternatively, cities can charge per bag for garbage pickup. This approach requires more extensive policy resources. In theory, per-bag pricing should encourage households to alter their consumption patterns—picking goods with less packaging, for example, or choosing cloth over disposable diapers. However, it can also have less desirable effects. For example, some people may engage in "midnight dumping" or the "Seattle stomp" of overstuffing individual bags of garbage. Don Fullerton and Thomas Kinnaman studied the implementation of per-bag pricing in Charlottesville, Virginia, and found that both recycling and illegal dumping increased.[53]

Of course, it remains an open question whether the underlying strategy of moving garbage from New York City to rural Virginia is a good thing. Does this transaction leave Virginia residents better or worse off?[54] A pessimist might wonder whether they are aware of what they are importing and what will become of the money the local government receives from New York City. An optimist would counter that due to the rural area's lower population density, fewer people will be exposed to the same amount of waste in Virginia than in New York. But the latter argument apparently carried little weight with Virginia governor Jim Gilmore, who wrote to Mayor Rudy Giuliani in 2001 to complain about the city's practice of exporting its trash. "I understand the problem New York faces," he noted. "But the home state of Washington, Jefferson and Madison has no intention of becoming New York's dumping ground."[55]

Furthermore, relying on market prices to create conservation incentives only works if land is private property. If a growing city wants to use someone else's land as a garbage dump, then it must pay compensation. But not all land is privately held. Suppose that a legal alternative to

---

53. Fullerton and Kinnaman (1996).

54. As chief economist at the World Bank, Larry Summers in 1991 discussed the same issues in the context of dirty trade between rich nations and poor nations. For a complete description of his memo, see "A World Bank Memo," *Baltimore Chronicle*, June 30, 1999 (baltimorechronicle.com/world_bank_jul99.html [April 2006]).

55. Lester R. Brown, "New York: Garbage Capital of the World," *Earth Policy Institute Update*, no. 10, April 17, 2002 (www.earth-policy.org/Updates/Update10_printable.htm [April 2006]).

dumping garbage at Fresh Kills is simply to dump the garbage in the Atlantic Ocean. Since nobody owns the ocean, the city would not face any demands for compensation. In cases like this, which exemplify the tragedy of the commons, rising incomes are unlikely to cause urban footprints to shrink.

Urban Land Management

Managing pollution is not the only environmental challenge that growing cities face. Urban land use policies also have an important impact on environmental sustainability. By preserving open space at the suburban fringe, zoning to minimize negative externalities, and repairing past blights, local leaders can help create greener cities. Richer governments are playing an active role on all three fronts.

PRESERVING OPEN SPACE. Throughout the United States, municipalities are purchasing open space around their borders to guarantee that the land is not developed. For example, the city of Boulder, Colorado, has earmarked a 0.73 percent sales tax to fund the purchase of 25,000 acres to establish a greenbelt around the city. It has also set aside 8,000 acres in the Boulder foothills to be used as parks. Some of the Boulder open space is leased to farmers and remains in agricultural use. Other parcels are maintained as natural areas. This allows residents to enjoy recreational activities such as walking, bicycling, and horseback riding. In the Seattle metropolitan area, King County has adopted a different strategy with a similar goal. Drawing upon a $50 million bond issue, the county is purchasing development rights from farmers. Farmers gain an increase in their income and in return they promise not to convert their "green space" into suburbia.

Such government initiatives solve a free-rider problem. In the absence of government intervention, environmental organizations such as land trusts might go door to door, asking people to contribute money to help preserve open space. But few people are likely to give under these conditions. The "win-win" for any one household is to contribute nothing to such programs and let everyone else underwrite their cost. As a result too little money is invested in protecting local public goods. Government's unique ability to collect taxes and allocate revenue solves this problem. However, not all governments can take this approach: like many green policies, "open space" initiatives are more likely to succeed as local incomes rise. After studying voting patterns for all open space referenda in the United States between 1998 and 2003, Matthew Kotchen and Shawn Powers found that richer jurisdictions and jurisdictions with more homeowners were more likely to vote to hold such ballot initiatives.[56] Once an open space referendum was put on the ballot, richer jurisdictions were also more likely to vote in favor of it.

GREENING THROUGH ZONING. Zoning also plays a key role in keeping a city green, largely by separating dirty production activities from areas where people live. Effective zoning is more likely to take place in richer cities with access to accurate land inventories and the ability to enforce local regulations. By contrast, municipalities in developing countries may have little information about which activities are taking place where and little ability to reorganize the landscape in a safe and efficient manner.

The results can be fatal, as a tragic episode from New York City's history shows. In 1911 the Triangle Shirtwaist Fire claimed 146 lives, in part because fire trucks were unable to make their way to a blazing factory through the neighborhood's narrow streets. This catalytic event triggered widespread interest in zoning in New York City. The adoption of the country's first zoning resolution soon followed in 1916.[57] This law divided New York City into use districts, area districts, and height districts. It also created four zoning classes designed to separate different types of economic activity: residential, business, retail, and unrestricted.[58] Taken together, these restrictions profoundly affected New York architecture and set a standard that was followed by many other cities.

CLEANING UP THE PAST. Every city has a history of past land use decisions that helps shape land use today. A city with more resources can allocate this money to preserve past "good moves" and to erase past mistakes. In some cases, such as cobblestone streets and Beaux Arts buildings, the heritage of the past is wonderful and should be protected. In other cases mistakes from the past live on. Boston's Big Dig is a classic example. When Boston's elevated highways were built in the 1950s, did Mayor John Hynes ever imagine that over $14 billion would eventu-

---

56. Kotchen and Powers (forthcoming). Nearly 1,000 jurisdictions had open space referenda, and nearly 80 percent passed. The Trust for Public Lands publishes the outcomes of these referenda in an annual survey called the Landvote.
57. Julia Palmer, "Letter to the Editor," *New York Times*, March 29, 2003, p. 24.
58. See "Zoning" (www.bartleby.com/65/zo/zoning.html [April 2006]).

ally be spent on the Big Dig project to knock them down? In the 1950, local politicians were primarily interested in faster highway access to the suburbs. Today, there is great demand to reconnect the downtown of Boston with the waterfront.

A prime example of the value—and cost—of cleaning up the past is provided by brownfields and Superfund sites. These sites are often the remains of defunct chemical companies. Their lasting legacy is contaminated land, which may potentially expose nearby residents to higher cancer risks. In many cases the polluting action was taken decades before the environmental damage was diagnosed. Such latency makes it difficult to hold the polluters accountable and forces the government to pick up the cleanup tab. These cleanups have been extremely costly, and even in their wake, many people remain concerned that the government itself does not know "how clean is clean."

Many firms avoid locating near such sites, fearing that they might one day be held liable for the consequences of past environmental damage. Rather than face such risks, many businesses prefer to settle in relatively pristine suburban locations. Recognizing this problem, in 2002 President George W. Bush signed into law the Small Business Liability Relief and Brownfields Revitalization Act, the first substantial set of amendments to the Comprehensive Environmental Response, Compensation, and Liability Act (CERCLA or Superfund) in years. This law was intended to ease "innocent" landowners' concerns about potential ex-post liability for contaminated sites.[59]

In addition, many cities have recognized the importance of cleaning up contaminated sites. This is particularly a priority for cities seeking to trade in their industrial past for a new tourist- and consumer-friendly reputation. For example, the city of Waukegan has invested millions in cleaning up Waukegan Harbor, which has been listed as one of the forty-three most polluted sites in the Great Lakes. By collecting contaminated

---

59. "Prior to these amendments, the innocent landowner defense—available to owners that did not know and had no reason to know of contamination upon acquisition—was so narrow that courts rarely accepted it. The amendments clarify with new detail that to be an innocent owner one must carry out 'all appropriate inquiries' prior to acquisition and take 'reasonable steps' to stop any new or continuing release. . . . Prior to the amendments, persons acquiring previously contaminated property, but who did not qualify as innocent landowners, were liable for cleanup of contamination regardless of the extent of their knowledge or care." See "CERCLA Amended to Limit Liability, Stimulate Brownfields Redevelopment" (library.findlaw.com/2002/Mar/8/132456.html [April 2006]).

materials and dredging the harbor, the city hopes to transform the waterfront from a jumble of brownfields and industrial plants into a regional destination with homes, shops, and restaurants.

With Superfund the government has created a pot of money to help clean up such sites. By 2000 some $30 billion had gone into cleaning up Superfund sites.[60] However, the effectiveness of this program has been limited. By early 1997, sixteen years after Congress enacted the Superfund legislation, only 11 percent of the sites on the Superfund National Priority List were clean.[61] Moreover, there is controversy surrounding the Environmental Protection Agency's approach to prioritizing sites. According to Hilary Sigman, the EPA has not targeted Superfund resources on the sites that pose the greatest threat to local quality of life. Instead, the evidence shows that sites in communities with higher voter turnout receive faster cleanup and that richer communities are more likely to appear on the National Priority List.[62]

Conclusion

Over the past thirty years, national and local regulation has played a major role in improving urban environmental quality in U.S. cities. In addition to voter support, such measures often require significant resources, which helps explain the shape of the urban EKC. Effective regulation is costly. Most firms and individuals will only alter their behavior if they believe that there are credible probabilities of being caught ille-

---

60. Greenstone and Gallagher (2005) examine home price growth in the areas surrounding 400 hazardous waste sites to be cleaned up through the Superfund program and compare their appreciation trends to home price growth in areas surrounding 290 hazardous waste sites that narrowly missed being cleaned up. This second set of areas represents a control group for the areas "treated" with a Superfund cleanup. These authors cannot reject the hypothesis that cleanups had no effect on home price growth. They conclude that the nearby home price appreciation benefits accrued due to Superfund cleanups are much lower than the average $43 million cost of a cleanup.

61. Sigman (2001).

62. Sigman (2001). Researchers have examined whether cleaning up noxious facilities helps to bring about environmental justice. Baden and Coursey (2002) study Chicago residents' proximity to noxious facilities between 1960 and 1990 and make the interesting observation that noxious facilities tend to be related to past employment centers. Whites wanted to live in these communities because they offered short commutes to high-paying jobs. Residential segregation against blacks actually kept this group away from these employment zones. An unintended consequence of this segregation was to reduce black exposure to noxious facilities in Chicago.

gally polluting and significant penalties if caught. Consequently, to avoid becoming "paper tigers," environmental regulators must invest in the ability to monitor millions of potential polluters every day.

The resulting data can do more than help regulators enforce the law; they can also give individuals the information they need to make better locational choices and help activists target their efforts. In any city with millions of economic agents, there are significant asymmetries of information between polluters and victims. Polluters, such as industrial plants, know what they have emitted, while nearby residential communities may have no idea. Policymakers can help level the playing field by creating information sources such as the EPA's Toxic Release Inventory (TRI). Created in the late 1980s, this publicly available database provides information by zip code on total emissions into the air, land, and water by local factories. Annual TRI announcements identifying the most polluting factories can be a public relations disaster for targeted firms. This creates an incentive for manufacturing plants to take costly steps to reduce their emissions rather than face a local community up in arms.

In ways such as this, environmental regulation has offered impressive public health and amenity benefits. But it has also imposed costs. In the 1970s and 1980s, there was extensive debate over regulation's impact on productivity growth. Today, it is still not known how much regulation increases prices for products such as cars.[63] Some researchers claim that car prices may be $2,000 higher because of Clean Air Act regulation whereas others argue that this figure overstates the true price increase because regulation has also increased vehicle quality.[64] In addition, economists continue to debate how regulation has affected firm profits and employment in the United States.[65] If these costs are substantial, richer nations are more likely to be willing to pay them to gain the benefits described in this chapter. This increased willingness also helps account for the shape of the EKC.

---

63. Hazilla and Kopp (1990).
64. Bresnahan and Yao (1985).
65. Goodstein (1999).

CHAPTER 6

Population Growth and the Urban Environment

The environmental Kuznets curve is based on the relationship between per capita income growth and the environment. The mechanisms that drive this relationship are the subject of the previous two chapters. But other varieties of urban growth—notably population and spatial growth–also help determine local environmental quality. This chapter focuses on how population growth affects urban "greenness," particularly in developing countries where it is commonly accompanied by increasing population density in urban areas.

Urban population growth is a key driver of environmental degradation. As more people crowd into cities, the problems of urban air pollution, water pollution, and solid waste production all grow worse. New migrants do not simply increase the scale of economic activity; they also tax and sometimes overwhelm basic infrastructure services. As a result urban sustainability, as measured by both the ecological footprint and public health approaches, declines. This is an ironic outcome, given that most new urbanites are drawn to cities by the hope of a better quality of life.

Urban Population Growth around the World

Around the world more people are living in cities. But are big cities becoming even bigger, or is the number of small and medium-size cities on the rise? Table 6-1 presents summary data on how city populations evolved in 143 countries between 1960 and 2000. The third and fourth

Table 6-1. *Change in Urban Population, 1960 and 2000*
Units as indicated

Nation	No. of cities	Average population exposure (millions of people)		Percent of urbanites living in a city of more than 2 million	
		1960	2000	1960	2000
Afghanistan	5	0.207	1.996	0	74.68
Albania	1	0.157	0.299	0	0
Angola	4	0.163	2.225	0	85.71
Argentina	14	4.686	8.265	67.35	63.14
Australia	11	1.435	2.344	34.31	54.29
Austria	6	1.234	1.448	0	66.09
Azerbaijan	2	0.886	1.719	0	0
Bahrain	1	0.056	0.146	0	0
Bangladesh	26	0.329	7.539	0	71.26
Belarus	5	0.345	1.148	0	0
Belgium	5	0.826	0.727	0	0
Benin	4	0.063	0.480	0	0
Bolivia	7	0.219	0.991	0	0
Bosnia	2	0.139	0.298	0	0
Botswana	1	0.004	0.213	0	0
Brazil	58	2.633	6.963	49.30	64.40
Bulgaria	7	0.413	0.703	0	0
Burkina Faso	2	0.056	0.936	0	0
Burundi	1	0.067	0.315	0	0
Cambodia	1	0.389	0.984	0	0
Cameroon	4	0.130	1.364	0	0
Canada	17	1.103	2.462	26.11	58.57
Central African Republic	1	0.080	0.636	0	0
Chad	1	0.076	1.043	0	0
Chile	22	1.201	3.328	56.74	58.46
China	113	2.306	3.590	33.28	48.13
Colombia	14	0.630	3.318	0	64.32
Congo, Republic of the	2	0.212	1.022	0	0
Costa Rica	1	0.283	0.988	0	0
Côte d'Ivoire	7	0.104	2.431	0	70.54
Croatia	2	0.367	0.948	0	0
Cuba	10	0.883	1.278	0	49.64
Czech Republic	4	0.745	0.849	0	0
Djibouti	1	0.041	0.503	0	0
Denmark	3	1.165	1.140	0	0
Dominican Republic	8	0.312	2.627	0	61.31
Egypt	15	2.320	6.740	52.75	75.34
El Salvador	1	0.247	1.408	0	0
Ecuador	10	0.296	1.594	0	42.61
Estonia	1	0.328	0.404	0	0
Ethiopia	3	0.459	2.383	0	89.69
Fiji	1	0.037	0.175	0	0
Finland	3	0.343	0.930	0	0
France	29	3.733	4.567	48.51	43.48
French Polynesia	1	0.028	0.106	0	0
Gambia	1	0.028	0.200	0	0
Germany	44	2.598	2.838	59.97	60.26
Ghana	3	0.314	1.526	0	0
Greece	5	1.836	2.395	79.30	71.77

continued on next page

Table 6-1. *Change in Urban Population, 1960 and 2000 (continued)*
Units as indicated

Nation	No. of cities	Average population exposure (millions of people)		Percent of urbanites living in a city of more than 2 million	
		1960	2000	1960	2000
Guatemala	1	0.532	3.242	0	100
Guinea	1	0.113	1.824	0	0
Guinea Bissau	1	0.018	0.280	0	0
Guyana	1	0.073	0.224	0	0
Honduras	3	0.104	0.794	0	0
Hong Kong	1	2.615	6.927	100	100
Hungary	5	1.453	1.358	0	0
Iceland	1	0.079	0.171	0	0
India	130	1.660	5.637	29.79	52.16
Indonesia	30	1.081	4.982	28.25	56.24
Iran	41	0.841	2.993	0	50.75
Iraq	12	0.519	2.443	0	42.83
Ireland	2	0.580	0.858	0	0
Israel	5	0.520	1.540	0	62.88
Italy	30	2.208	2.140	54.39	54.67
Jamaica	1	0.377	0.913	0	0
Japan	156	3.916	9.909	39.44	49.84
Jordan	3	0.164	1.082	0	0
Kazakhstan	11	0.294	0.596	0	0
Kenya	6	0.170	1.513	0	57.54
Kyrgyzstan	2	0.299	0.669	0	0
Kuwait	3	0.192	0.890	0	0
Latvia	1	0.597	0.775	0	0
Lebanon	4	0.446	1.705	0	81.45
Lesotho	1	0.006	0.165	0	0
Liberia	1	0.041	1.348	0	0
Lithuania	3	0.253	0.458	0	0
Madagascar	5	0.178	1.130	0	0
Malawi	2	0.077	0.497	0	0
Malaysia	15	0.178	0.621	0	0
Mali	8	0.176	0.769	0	0
Martinique	1	0.085	0.095	0	0
Mexico	142	1.998	6.394	34.28	43.40
Mongolia	1	0.195	0.738	0	0
Morocco	10	0.500	1.952	0	40.62
Mozambique	6	0.113	2.222	0	70.40
Myanmar	13	0.576	2.704	0	59.60
Namibia	1	0.036	0.192	0	0
Nepal	4	0.080	0.492	0	0
Netherlands	20	0.491	0.604	0	0
New Zealand	6	0.273	0.721	0	0
Nicaragua	5	0.142	0.688	0	0
Niger	4	0.024	0.340	0	0
Nigeria	67	0.238	5.265	0	42.75
Norway	3	0.447	0.761	0	0
Pakistan	47	0.824	5.353	0	65.26
Panama	1	0.283	1.173	0	0
Papua New Guinea	1	0.014	0.300	0	0
Paraguay	1	0.309	1.262	0	0

continued on next page

Table 6-1. *Change in Urban Population, 1960 and 2000 (continued)*
Units as indicated

Nation	No. of cities	Average population exposure (millions of people)		Percent of urbanites living in a city of more than 2 million	
		1960	2000	1960	2000
Peru	16	1.157	4.898	0	63.86
Philippines	62	0.896	4.600	35.92	46.39
Poland	27	1.160	1.579	28.58	42.56
Portugal	5	0.700	3.027	0	63.06
Puerto Rico	2	0.473	1.257	0	0
Reunion	1	0.066	0.132	0	0
Romania	20	0.693	0.835	0	31.93
Russia	111	1.746	2.436	28.12	24.63
Rwanda	1	0.004	0.351	0	0
Sahara, Western	1	0.005	0.183	0	0
Saudi Arabia	5	0.140	2.574	0	73.80
Senegal	5	0.252	1.551	0	71.68
Sierra Leone	3	0.110	0.857	0	0
Singapore	1	1.634	3.567	0	100
Slovakia	1	0.242	0.449	0	0
Somalia	4	0.068	0.894	0	0
South Africa	31	0.568	1.436	0	36
South Korea	30	1.115	4.373	33.72	60.66
Spain	33	1.106	1.841	24.39	46.14
Sri Lanka	6	0.325	0.465	0	0
Sudan	7	0.210	1.808	0	47.23
Suriname	1	0.123	0.214	0	0
Sweden	12	0.443	0.887	0	0
Switzerland	3	0.416	0.743	0	0
Syria	10	0.361	1.245	0	0
Tajikistan	1	0.312	0.692	0	0
Tanzania	10	0.095	1.532	0	35.94
Thailand	1	2.151	7.281	100	100
Tunisia	6	0.426	1.416	0	0
Turkey	36	0.797	4.240	0	56.79
Turkmenistan	1	0.170	0.605	0	0
Uganda	1	0.137	1.212	0	0
Ukraine	38	0.520	0.988	0	15.55
United Arab Emirate	2	0.048	0.886	0	0
United Kingdom	41	4.058	3.236	55.95	52.84
United States	170	3.468	4.035	36.78	48.32
Uruguay	1	1.155	1.236	0	0
Uzbekistan	8	0.601	1.267	0	52.07
Venezuela	34	0.582	1.413	0	21.73
Vietnam	12	0.872	3.224	0	69.82
Yemen	3	0.115	1.301	0	0
Yugoslavia	2	0.564	1.339	0	0
Zaire	9	0.250	3.142	0	56.53
Zambia	8	0.079	1.004	0	0
Zimbabwe	5	0.141	1.196	0	0

Source: Based on a data set created by Vernon Henderson for the years 1960, 1970, 1980, 1990, and 2000. Only values for the 2,096 cities for which data were available in both 1960 and 2000 are used. See J. Vernon Henderson, "World Cities Data" (www.econ.brown.edu/faculty/henderson/worldcities.html [September 2002]); Henderson and Wang (2004).

columns of this table show the population exposure of the average urbanite in those two years, calculated as a weighted average. For example, if a nation has three cities and one has 2 million people and the other two have a population of 1 million each, then the average urbanite in this nation lives in a city with 1.5 million people: (2*0.5) + (1*0.25) + (1*0.25). The last two columns show the percentage of urbanites living in a city of at least 2 million people.

The average population exposure in a particular year can be calculated as follows:

$$\Sigma\ s(jt)^*\text{Pop}(j,t),$$

where $s(jt)$ represents the share of the world's population that live in city j in year t. In any given year, the shares sum to one. $\text{Pop}(j,t)$ represents the population level of city j in year t. Based on this formula, in 1960 the average urbanite lived in a city with 2,232,847 people. The corresponding figure for 2000 was 4,247,185 people. This represents an annual growth rate of 2.3 percent. However, this world average masks important heterogeneity: in Brazil average urban population exposure grew by more than 160 percent over this period, while in Italy the same measure fell by 3 percent. The same heterogeneity can be observed in the percentage of urbanites living in large cities. In Afghanistan, for example, the proportion of urbanites living in cities with at least 2 million people grew from 0 percent to 75 percent between 1960 and 2000. But in a number of developed countries, including France and the United Kingdom, the corresponding figure fell. Globally, 42 percent of urbanites lived in cities with over 2 million people in 1960, and by 2000 this percentage had grown to 48 percent.

Much of this growth occurred in more tropical areas. In 1960 the average urbanite lived at a latitude of 31.31 degrees north—roughly the location of Shanghai. But in 2000 the average urbanite lived roughly six degrees further south, around the same latitude as Delhi, Havana, and Karachi. In addition, cities have been growing more rapidly in Asia and Africa than in Europe or the Americas. Table 6-2 shows that the population of the median African city grew by 227 percent over this period, while the median European city's population grew by only 42.3 percent.

In analyzing these data, Vernon Henderson and Hyoung Gun Wang find that concern about the growth of "megacities" has been exaggerated.[1] In 2000 there were nineteen cities around the world with a popula-

---

1. Henderson and Wang (2004).

Table 6-2. *Urban Population Growth by Continent, 1960–2000*[a]
Units as indicated

	Percent change in city population growth			
Percentile	Africa	Americas	Asia	Europe
1st	–79.5	–8.6	–19.9	–33.4
5th	26.7	10.1	20.1	–14.5
10th	38.3	20.6	38.0	–6.5
25th	105.4	64.7	77.5	8.1
50th	227.2	142.3	160.9	42.3
75th	498.0	283.6	278.9	82.3
90th	914.9	416.3	443.0	135.7
95th	1,348.8	576.3	683.9	186.6
99th	1,736.1	958.0	1,300.0	443.8
Summary statistic				
Average	390.0	199.0	226.0	61.9
No. of cities	81	244	475	364

Source: Author's calculations based on data from Henderson and Wang (2004).

a. This table reports the empirical distribution of the percent change in city population growth between 1960 and 2000 for those cities that had more than 100,000 people in the year 1960.

tion of at least 10 million, but these cities contained only 14 percent of the world's urbanized population. Clearly, a majority of urbanites do not live in megacities. From an environmental perspective, is this good news?

Megacities do offer some sustainability advantages. Most important, they can capitalize on economies of scale in developing "green" investments, such as public transit, sewers, and water systems. Such projects require enormous upfront investment, but the marginal cost of providing service to recent migrants can be quite low. Nations that concentrate residents in megacities can reduce the average cost of providing such basic services to residents relative to nations where the population is scattered across many smaller cities.

In addition, megacities offer diverse local labor markets. While smaller cities, such as Gloversville, New York, typically specialize in one industry (gloves, in this case), major cities represent diversified local economies. This facilitates women's labor force participation, which in turn tends to result in smaller families.[2] At the national level, there is a clear negative relationship between urbanization and subsequent popu-

---

2. When a married couple moves to a smaller city, the wife tends to face problems in finding a good job (Ofek and Merrill 1997; Costa and Kahn 2000). In a major city, such colocation problems are less likely to arise.

lation growth. Using cross-national data for 110 countries, the correlation between a nation's 1965 share of total population who live in cities with the nation's percent change in population growth between 1965 and 1995 is −0.64.[3] Many environmentalists believe that restraining population growth is essential for achieving long-run sustainability, whether at the urban or the national level.[4]

Despite these advantages, there are important public health risks associated with megacity growth. Such cities are more susceptible to epidemics, for example. When the Black Death struck Europe in the fourteenth century, bigger cities had more plague outbreaks. A cliché from portfolio theory is that wise investors do not put all of their eggs in one basket, but megacity urbanization does just that. In the United States, even enormous metropolitan areas such as Los Angeles and New York City contain less than 5 percent of the nation's population. In other nations, such as Argentina, 30 percent of the nation's population lives in the country's biggest city. Such concentration could pose major public health challenges if an event such as a waterborne epidemic occurred.

In addition, when the urban population is concentrated in one megacity or a small number of megacities, it becomes much more difficult for residents to "vote with their feet." Particularly in developed nations, decentralized competition between cities creates an incentive for politicians to adopt green policies, as discussed in the previous chapter. In this sense the existence of several large cities represents a type of insurance policy for urbanites. In a nation with hundreds of cities, if one city suffers from declining environmental quality due to population growth, its own residents will exit, and potential entrants will choose to live somewhere else. This out-migration will mitigate the city's original sustainability challenge.

Whether a nation's urban population is concentrated in a megacity depends largely on the policies the nation adopts. Urban economists have used cross-national data to identify the key determinants of city size. According to Alberto Ades and Edward Glaeser, the population of a nation's major city will be larger if any of the following conditions holds: the city is a capital city, the nation is a dictatorship, the nation is in Latin America, and the nation's economy is closed to international trade.[5] This last empirical finding is important because more nations are

3. See Arzaghi and Henderson (2005).
4. See Cohen (1995).
5. Ades and Glaeser (1995).

Figure 6-1. *Ambient Particulate Levels and City Population, 1995*

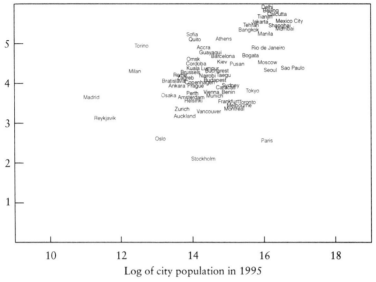

Source: Author's calculations based on data from the World Bank (2001).

opening up their economies to trade. If trade openness causes a nation's urban population to spread across many cities, one benefit could be important gains in urban environmental quality.

Population Growth and Urban Environmental Quality

While megacities may face the greatest sustainability challenges, smaller cities can also see environmental quality suffer as they grow. The following discussion briefly examines how population growth affects three of the leading environmental problems cities face: air pollution, water pollution, and the management of solid waste. It explores how migration to cities is likely to affect the ecological footprint of a nation or of the world.

Air Pollution

By and large, ambient air pollution worsens as city populations grow. Figures 6-1 and 6-2 graph levels of ambient particulates and sulfur diox-

Figure 6-2. *Ambient Sulfur Dioxide Levels and City Population, 1995*

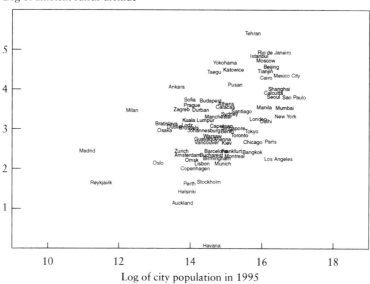

Source: Author's calculations based on data from the World Bank (2001).

ide against city size, based on World Bank data from 1995.[6] There is clearly a positive correlation between levels of particulates and sulfur dioxide, on the one hand, and city size, on the other. These figures provide a cross-city snapshot, but the ideal would be to observe how urban air quality evolves over time for the same city as it grows. Unfortunately, it is quite difficult to collect the necessary data over time, especially for cities in developing countries.

It is also important to note that these simple figures focus solely on cross-city differences in population. World Bank researchers have documented that in addition to a city's population level, its climate conditions, governance, and income all play important roles in determining local air pollution levels.[7] This point is reinforced by evidence from the United States. The Los Angeles basin suffers from the highest levels of air pollution in the United States. But Los Angeles has made dramatic progress on air pollution over the last twenty-five years. For ambient ozone, the main

6. World Bank (2001).
7. Dasgupta and others (2004).

ingredient of smog, the number of days per year exceeding the federal one-hour ozone standard declined from about 150 at the worst locations during the early 1980s to 20 to 30 days per year as of 2005. Recent gains against pollution are especially notable because Los Angeles County's population grew by 29 percent between 1980 and 2000.[8] This example highlights how technological improvements can offset population growth enough that local environmental progress takes place.

The U.S. Environmental Protection Agency's Annual Summary Table Query database makes it possible to examine the relationship between population growth and ambient pollution levels.[9] The Environmental Protection Agency widely monitors air pollution, and most of these monitoring stations are located in relatively heavily populated counties. Table 6-3 reports five different regression estimates based on data collected at these sites. For example, consider the results for ambient total suspended particulates. Controlling for a monitoring station fixed effect, which absorbs time-invariant factors such as geography, average wind speeds, and climate, I find that a 10 percent increase in a county's population increases ambient particulate levels by 4.4 percent. However, when county population is held constant, ambient particulates decline by 2.6 percent each year, presumably due to regulation and the phasing-in of cleaner technology. In other words, urban population growth does raise county air pollution levels, but in this case technological progress largely offsets this effect.[10]

Water Pollution

Urban population growth can also overwhelm local efforts to provide key services, such as clean water. As poor migrants enter a city, they increase the demand for basic services but are often incapable of contributing financially to their supply. As existing services become over-

---

8. See Bureau of Economic Analysis, "Regional Economic Accounts—Local Area Personal Income" (www.bea.gov/bea/regional/reis [May 2006]).

9. See U.S. Environmental Protection Agency, "Monitor Data Queries: Annual Summary Table Query" (www.epa.gov/aqspubl1/annual_summary.html [April 2006]).

10. For example, consider the sulfur dioxide results reported in table 6-3. Suppose that a county's population grew by 10 percent in one year. This would be enormous growth. The regression results indicate that ambient sulfur dioxide levels would increase by 3.7 percent, but the coefficient on the time trend equals –0.037. Thus, on average, this county would experience *no change* in ambient pollution because the time trend effect would just offset the pollution consequences of the population growth.

Table 6-3. *U.S. Air Pollution Dynamics and Population Growth, 1973–2000*[a]

Independent variable	Dependent variable				
	Log(total suspended particulates)	Log(carbon monoxide)	Log(sulfur dioxide)	Log(nitrogen dioxide)	Log(ozone)
Log(county population)	0.435	0.514	0.374	0.085	0.043
	(0.013)	(0.029)	(0.040)	(0.035)	(0.013)
Time trend	–0.026	–0.050	–0.037	–0.013	–0.001
	(0)	(0)	(0.001)	(0.001)	(0.000)
Constant	–0.987	–5.921	–9.344	–5.010	–3.546
	(0.154)	(0.382)	(0.480)	(0.434)	(0.157)
Monitoring station fixed effect	Yes	Yes	Yes	Yes	Yes
Summary statistic					
Adjusted R^2	0.813	0.865	0.809	0.862	0.564
Number of observations	71,344	14,157	34,448	19,184	45,533

Source: Author's calculations based on pollution data from U.S. Environmental Protection Agency, "Monitor Data Queries: Annual Summary Table Query" (www.epa.gov/aqspubl1/annual_summary.html [April 2006]).

a. Each column of this table reports a separate fixed effects regression using data from 1973 to 2000. The unit of analysis is a monitoring station. The dependent variable is based on the annual arithmetic mean at a monitoring station in a specific year. Standard errors are shown in parentheses. The explanatory variable called log(county population) is based on the population of the county where the monitoring station is located.

taxed, their quality falls. This can have serious public health consequences. Poor levels of sanitation are associated with more than fifty communicable diseases including diarrhea, ascariasis, trachoma, hookworm, schistosomiasis, and guinea worm.[11] Conversely, improved sanitation can deliver important public health gains. In Brazil during the 1980s, after efforts to increase urban access to higher-quality water, deaths caused by waterborne diseases fell by 50 percent for children under age fifteen.[12]

In theory the capital markets should be able to respond to growing cities' needs. There are investors on Wall Street with the capital to build water infrastructure systems for growing cities in developing countries. In such cities there are millions of people who want higher-quality water—and will be able to pay for it over time as their incomes increase. All parties would be better off if Wall Street provided the upfront invest-

11. Beede and Bloom (1995).
12. Seroa da Motta and Rezende (1999).

ment necessary to provide basic services and if local governments pledged to pay them back out of future revenues. Unfortunately, a problem arises because the water system investment is irreversible. Investors know that once the water system is built, they can only recover their investment if water prices remain relatively high. They also know that once their access to water is ensured, the urban poor will have an incentive to lobby for lower prices—and that ambitious politicians will have an incentive to give in. As a result investors do not invest, and clean water remains a distant goal.[13]

This problem of "ex-post" renegotiation is most serious in the poorest urban neighborhoods. If municipal authorities do not believe that people will pay their bills, they are unlikely to invest in connecting them to city systems. As a result a city's poorest residents often end up paying the most for water and other basic services—or do without. In Brazil's shantytowns residents pay ten times the legal rate to water pirates who tap illegally into the main systems.[14] In addition, in such areas large amounts of sewage and other waste are often dumped into local waters. In the past, when cities had much smaller populations, this may have been an effective waste disposal solution. Natural capital, such as cold, deep, fast-moving water, worked to take care of the waste problem. But in fast-growing cities, the absorptive capacity of such natural capital is often overwhelmed.

The resulting pollution problems are likely to be particularly acute when downstream cities or countries must cope with the resulting mess. Recent research has focused on measuring the extent to which pollution increases when at least some of the consequences are borne by another nation. After studying water pollution levels based on biochemical oxygen demand, which rises in response to human activities such as the dumping of sewage, Hilary Sigman finds that pollution levels are 40 percent higher at international border monitoring stations than at in-country monitoring stations.[15]

Solid Waste

Growing cities, not surprisingly, produce more garbage, and in the absence of significant income growth, they often lack the means to deal

13. Noll, Shirley, and Cowan (2000).
14. Katakura and Bakalian (1998).
15. Sigman (2002)

with it effectively. Residents are particularly sensitive to this problem, as a recent report from the International Development Bank observes:

> When asking people in the streets of the smaller cities in Latin America about the most important problems of the city, there is a fair chance that they mention "la basura" (solid waste) as the most important problem. In part, this has to do with the fact that it is a very visible problem. In many cities, waste is scattered in streets, squares, parks, water flows, etc., because collection systems are absent or inappropriate. Also, the relationship with health problems is clear, because of the obvious presence of disease vectors like flies, rats, dogs and others that feed and breed in the waste heaps.[16]

Survey research in Bangkok, Thailand, has indicated similar patterns. Respondents prioritized garbage problems and resulting rodent populations as the city's major environmental hazard.[17]

The environmental impact of solid waste production is most severe in cities where property rights are not enforced. In a city with well-established property rights, residents would be forced to pay compensation if they disposed of their trash on someone else's property, including property belonging to the state. Absent this expectation, people are much more likely to dump their garbage in public streets or parks. One possible solution to this problem, which has been put forward by Hernando de Soto, is to give squatters and other informal residents formal property rights to their land.[18] This would provide them with a legal asset that they could trade and borrow against. From a green cities perspective, it would also create a "stakeholder society" with stronger incentives to protect local environmental quality.[19] Using U.S. and German data, researchers have documented that homeowners are more engaged citizens than renters.[20] Such homeowners have a financial incentive to take personally costly actions, such as confronting neighbors, to preserve local quality of life because if they fail to act, the values of their homes may decline.

16. See Vries and others (2001).
17. Daniere and Takahashi (1997).
18. Soto (2000).
19. Hoy and Jimenez (1996).
20. DiPasquale and Glaeser (1999).

Urbanization and Consumption

Thus far, this chapter has examined how urban population growth affects local environmental indicators. These negative effects would be observed even if new urbanites lived just as they did in their former homes. But consumption patterns typically change as poor rural residents move to big cities and as immigrants move from cities in poorer countries to cities in developed nations. These migrants often end up earning and consuming more, which can be a win for them but a loss for the environment. For example, a Mexican immigrant to Los Angeles may earn enough to consume 500 gallons of gasoline a year even though he could not afford a car back home. Thus some environmentalists are worried about the consumption scale effects induced by migration to richer nations. As Jason DinAlt argues, "The last thing the world needs is more Americans. The world just cannot afford what Americans do to the earth, air, and water."[21] If this argument is taken seriously, it suggests that environmentalists should support limits on immigration to the United States in order to reduce the world's ecological footprint.[22]

Diversity and Growing Cities

Urban population growth does more than simply scale up city size. Typically, it also creates a more diverse urban population. Diversity is a defining characteristic of major cities. Diverse cities offer a far greater range of jobs, cultural opportunities, and even cuisine. Jane Jacobs's work celebrates diversity as a key engine of urban growth.[23] But diversity also has costs: more diverse cities are harder to govern because their residents are more likely to disagree over important public policy issues and often have very different goals. As Raymond Vernon once observed,

> If a major object of our existence were to create great cities of beauty and grace, there would be something to be said in favor of dictatorship. As a rule, the great cities of the past have been the

---

21. Jason DinAlt, "The Environmental Impact of Immigration into the United States" (www.carryingcapacity.org/DinAlt.htm [April 2006]).
22. The Sierra Club, for example, continues to wrestle with its policy stand on U.S. immigration. See Sierra Club, "Tale of Two Immigrants" (www.sierraclub.org/sierra/200411/immigrants.asp [November-December 2004]).
23. Jacobs (1969).

cities of the powerful city-states in which a dominant king or governing body had the power and the will to impose its land use strictures upon an obedient populace. Weak or divided local governments, responsive to the push and pressure of the heterogeneous interest groups which make up a city, have rarely managed to intervene enough to prevent the unpalatable kind of growth which typifies our larger urban areas.[24]

Public transit investments, to take one example, are less likely to have widespread support in diverse cities where people have very different types of jobs, schedules, and housing opportunities. Similarly, "green" pricing for basic services can be a casualty of heterogeneity—or more specifically, income inequality linked to urban population growth. As income inequality grows, it becomes increasingly difficult to implement pricing policies that generate adequate revenue and provide signals of resource scarcity without putting services out of reach for many residents. In extreme cases basic infrastructure projects, such as water and waste treatment plants, may never get built. In other instances local authorities may adopt simpler fixed-rate pricing schemes that give consumers no incentive to conserve on the resources they consume.

Data on water pricing illustrate a similar effect. In the United States, residential consumers are responsible for two-thirds of urban water use.[25] Moreover, a few uses account for most of this consumption: flushing the toilet (36 percent), baths and showers (28 percent), and laundry (20 percent). Increasing the price of water would reduce consumption in most households. However, policymakers frequently prefer to mandate technologies such as low-flow toilets. Raising the price of a necessity is an unpopular move, especially in areas where inequality is high and median incomes are low. Based on data from twenty-three California cities, Christopher Timmins has shown that water charges account for a smaller share of municipal revenues, compared to relatively progressive property and sales taxes, in cities with lower median incomes.[26] While equity considerations may justify this approach, cities that adopt it are less likely to see their ecological footprints shrink.

Achieving consensus for adopting green policies is even more challenging when urban population growth contributes to a rise in ethnic as

24. Vernon (1964, p. 97)
25. Timmins (2002).
26. Timmins (2002).

well as economic diversity. All cities suffer from anomie to some degree because most urbanites are strangers to one another. But cities that are fragmented along economic and ethnic lines typically have lower levels of social cohesion and social capital. In consequence, they have greater difficulty resolving collective action problems, such as reducing litter in public parks and public spaces.

Recent research has found that public goods provision is lower in more diverse communities.[27] For example, Edward Miguel and Mary Kay Gugerty find that there is less school funding and lower quality water well maintenance in more ethnically diverse communities in Kenya.[28] Other studies show that governments invest less in education, roads, and sewers in more ethnically fragmented areas.[29]

Diversity also appears to reduce support for redistribution. Peter Lindert documents that among member nations of the Organization for Economic Cooperation and Development, those with greater income inequality spend less on social programs.[30] Similarly, differences in ethnic heterogeneity help explain why the welfare state is considerably more generous in Europe than in the United States.[31] In the latter minorities are overrepresented among welfare recipients, which tends to reduce support for redistribution among many whites.[32] This is relevant for analyses of urban sustainability because many big city public projects, such as public transit expansion, implicitly redistribute income to the urban poor.

Finally, greater heterogeneity is associated with lower levels of civic engagement. For example, households in more ethnically homogeneous counties are more likely to fill out census forms whereas free riding on others' efforts is more common in cities where diversity is high.[33] In cities and communities with greater levels of racial heterogeneity and income inequality, people are less likely to volunteer and less likely to be members of civic groups. Such cities are more likely to be "brown" because residents litter and pollute without considering the greater social consequences of their actions. While any one individual's contri-

27. Alesina, Baqir, and Easterly (1999).
28. Miguel and Gugerty (2005).
29. Alesina and others (2003).
30. Lindert (1996).
31. Alesina and Glaeser (2004).
32. Luttmer (2001).
33. Vigdor (2004); Costa and Kahn (2003a).

bution to pollution may be quite small, in a city with millions of people, these actions add up to a large environmental impact.

Conclusion

Cities differ in their ability to absorb population growth without experiencing local environmental degradation. Some of the factors that affect the relationship between growth and sustainability are relatively immutable, such as climate and geography. But the quality of governmental institutions also plays a key role in determining whether a city will be able to cope with population growth. Long-term planning requires resources and expertise. Ideally, urban planners would be able to forecast likely urban population growth over the next twenty years. Anticipating this growth, city leaders would take proactive steps to limit its environmental impact by financing necessary infrastructure, such as roads, sewerage, and water treatment plants. Cities that are either unwilling or unable to take such steps will suffer greater environmental degradation as a result of urban growth.[34]

---

34. Cutler and Miller (2005).

CHAPTER 7

Spatial Growth: The Environmental Cost of Sprawl in the United States

Since the end of World War II, most of the growth in U.S. metropolitan areas has taken place in the suburbs.[1] In 1940, 48 percent of the U.S. population lived in a metropolitan area, and 68 percent of metropolitan area residents lived in center cities. By 1990 the first figure had grown to 78 percent, and the second had fallen to 40 percent.[2] In 1970 the average urbanite lived in a community with 10,452 people per square mile. By 2000 urban population density had fallen over 25 percent, with the average metropolitan area resident living at a density of 7,358 people per square mile.[3]

This chapter examines the environmental costs of this trend, which has come to be labeled "urban sprawl." Sprawl, which I define as the migration of homes and jobs to low-density areas, poses several sustainability challenges. It typically increases land consumption and vehicle use, which in turn increases carbon dioxide production and requires the building of new roads. In addition, sprawl increases the proportion of middle-class households that are likely to oppose policies, such as expanding mass transit, that improve urban sustainability. Instead, these voters have a strong incentive to support policies that subsidize private transportation.

1. Margo (1992); Mieszkowski and Mills (1993); Glaeser and Kahn (2004).
2. Altshuler and others (1999).
3. This calculation is based on all census tracts within twenty-five miles of a major central business district.

Explaining Sprawl

Over the last one hundred years, a variety of forces have contributed to urban sprawl. Transportation innovations have sharply reduced the total cost of commuting within metropolitan areas, as vehicles have become faster and cheaper in real terms.[4] This development has allowed workers to hold on to city center jobs while moving their families to the suburbs in search of a higher quality of life. Many middle-class households have migrated to the suburbs because they offer newer, more spacious homes than center cities, where a majority of the housing stock can be over seventy years old. They have also been drawn out of cities by the lure of better public schools and the desire to flee urban crime.[5] Many suburban communities, such as Scarsdale, New York, represent clubs of richer people willing to pay higher taxes for better services.[6]

For decades suburbanites had to endure long commutes in return for their higher quality of life. But more recently, jobs have migrated to the suburbs as well. New York City is one of the rare major metropolitan areas to retain a significant proportion of employment downtown. In 2000, 58 percent of the New York metropolitan area's employment was located within ten kilometers (roughly 6.2 miles) of the city's central business district. Boston was not far behind, at 52 percent, but in Dallas and Los Angeles, only 21 percent of the metropolitan area's employment was located within ten kilometers of downtown.[7]

Companies employing skilled professionals, such as law, accounting, and finance firms, used to cluster in central business districts such as Wall Street. This proximity facilitated deal making, reputation building, and learning from other firms. But telecommunications innovations, ranging from the telephone to fax machines to the Internet, now allow far-flung businesses to interact. As a result more corporations have

---

4. Jackson (1985); Glaeser and Kahn (2004).
5. Cullen and Levitt (1999). Over the last twenty years, crime has fallen in many major U.S. center cities (Levitt 2004). This relative improvement in center city quality of life will provide an important test of whether "flight from blight" is an important explanation for sprawl. If this theory is correct, sprawl should slow in those metropolitan areas where crime has decreased the most.
6. As wealthier households "vote with their feet" by moving to the suburbs, urban politicians face growing limits on their freedom of action. Suburbanization causes the tax base of center cities to decline. It also makes it more difficult for politicians to implement policies opposed by high-income residents, who now have more exit options.
7. Baum-Snow and Kahn (2005).

sought out suburban campuses where land is cheaper and taxes are relatively low. Perhaps not coincidentally, such moves also allow suburban executives to reduce their daily commutes, both by living closer to their jobs and by driving at higher speeds.[8] One example of a major company building a suburban corporate campus at the fringe of a big city is Microsoft's new operation in Richmond, Washington: it will be 10 million square feet in size after expansion is complete, and there will be 12,000 workers there.[9]

Employment sprawl in turn fuels further residential sprawl. Shorter commute times increase the appeal of the suburbs for highly educated professionals who are already spending too many hours on the job. And the suburbanization of such high-skill workers encourages others to follow in their wake. Soon, upscale shops and restaurants start to locate in suburban malls, further reducing demand to work, live, and shop in center cities.

Federal transportation policy has played an important role in encouraging these trends. A notable example is the Federal-Aid Highway Act of 1956, which funded the building of high-speed roads in the 1950s and 1960s. According to an analysis by Nathaniel Baum-Snow, the population living in center cities would have grown by 6 percent between 1950 and 1990 had the interstate highway system not been built.[10] Instead, between 1950 and 1990, the aggregate population of center cities in the United States declined by 16 percent, despite national population growth of 64 percent.

In addition, advances in technology, such as the development of air conditioning, have spurred migration to low-density Sun Belt cities, such as Las Vegas and Phoenix, that were mostly built up after the diffusion of the automobile. Between 1969 and 2002, Las Vegas's population grew by 173 percent, and Phoenix's grew by 124 percent, while the Detroit and Philadelphia metropolitan areas hardly grew at all. Using

8. Gordon, Kumar, and Richardson (1991).
9. Brier Dudley, "Huge Microsoft Expansion to Ripple through Region," *Seattle Times*, February 10, 2006 (seattletimes.nwsource.com/html/microsoft/2002796093_microsoft10.html).
10. Baum-Snow (2005). There is a certain irony that these highways had this effect because at the time they were built, they were thought of a way to save cities. According to Pietro Nivola, "Earlier in the century, the same cities were widely deemed overcrowded and clogged and that routing traffic out and around them would provide relief. To decongest their downtowns, city officials, merchants, and housing experts, not just developers of the suburban subdivisions and shopping centers, pleaded for highway bypasses." See Nivola (1999, p. 14).

data for 922 U.S. cities with more than 25,000 residents in 1980, I find that population growth between 1980 and 2000 was highest in warm winter, low-rain cities such as San Diego.[11]

Measuring Sprawl's Environmental Impact

Suburban growth can cause a number of environmental problems, including air pollution, greenhouse gas production, habitat destruction, increased water consumption, and the destruction of open space. This section now turns to measuring some specific environmental consequences of sprawl.

Increased Vehicle Use

Between 1950 and 1994, the number of miles driven annually in the United States increased by 140 percent, while the U.S. population increased by 50 percent.[12] Two major factors lie behind this upward trend. The first is simply income growth. Using data from the 1995 National Personal Transportation Survey for households who live in a metropolitan area, I find that a 10 percent rise in household income increases annual household vehicle miles by 14 percent.[13] But the second reason for the increase in driving is the growth of the suburbs. On average, suburban households drive 31 percent more miles per year than households of the same size and income that live in center cities.[14] Similarly, annual mileage differs between more and less spread-out cities. For example, the average Atlanta household would drive 25 percent fewer miles if it relocated to relatively compact Boston.[15] As a result there are significant differences in average gasoline consumption across the coun-

---

11. I run an ordinary least squares regression of each city's percent change in population between 1980 and 2000 as a function of the log of its 1980 population and the city's average January temperature and average annual rainfall. In this regression, R^2 equals 0.1599 and standard errors are reported in parentheses.

Growth = 0.8023 − 0.0612*log(population 1980)
(0.1852) (0.0164)

+ 0.0096*(January temperature) − 0.0064*rain
(0.0009)   (0.0009)

12. Nivola (1999).
13. Kahn (2000).
14. Kahn (2000).
15. Bento and others (2005).

try. Cross-national studies suggest that gasoline consumption could be 20 to 30 percent lower in sprawling cities like Houston and Phoenix if their urban structure more closely resembled that of Boston or Washington, D.C.[16]

I find similar results using data from the United States. Based on the Department of Transportation's *2001 National Household Travel Survey*, I calculate for forty-nine metropolitan areas how much gasoline the average household of four people with an income of $45,000 uses annually.[17] The ideal would be to observe the same household's gasoline consumption if it lived in every metropolitan area in the United States; this would provide evidence on which metropolitan areas are "most sustainable" based on this consumption criterion. To approximate this ideal, I standardize households using regression techniques to control for household income and household size. In particular, I estimate that

$$\text{Gasoline}_{ij} = \text{MSA}_j + B_1 * \log(\text{income } i) + B_2 * \log(\text{household size } i) + U_{ij}.$$

In this regression the dependent variable is the annual consumption of gasoline by household *i* residing in metropolitan area *j*. MSA represents the metropolitan statistical area, B_1 and B_2 stand for the regression coefficients, and U stands for the error term. Controlling for household size and household income, I estimate metropolitan area fixed effects (the MSA_j vector).

The results are reported in table 7-1. The five cities featuring the lowest gasoline consumption are New York, San Francisco, Sacramento, Pittsburgh, and Boston—all relatively compact, high-density cities. The ten cities with the highest predicted gasoline consumption include sprawling Atlanta, Charlotte, San Antonio, Orlando, and Houston. In Houston the average household of four people and $45,000 in income consumes 1,407 gallons of gasoline per year, but the same family would use only 919 gallons annually if it lived in San Francisco. These differentials are important because gasoline consumption is a major reason that the United States produces 20 percent of the world's greenhouse gas emissions.[18]

---

16. Newman and Kenworthy (1999).
17. See U.S. Department of Transportation, "National Household Travel Survey—Downloads" (nhts.ornl.gov/2001/html_files/download_directory.shtml [May 2006]).
18. Burning a gallon of gasoline releases nineteen pounds of carbon dioxide into the atmosphere. Returning to a key theme of chapter 2, a "green city" would offer

Table 7-1. *Average Annual Household Gasoline Consumption by Metropolitan Area, 2001*
Units as indicated

Metropolitan area	Predicted gallons per year	Metropolitan area	Predicted gallons per year
New York City	783.577	Kansas City	1,192.610
San Francisco	919.220	Hartford	1,200.886
Sacramento	969.901	Oklahoma City	1,212.776
Pittsburgh	993.004	Milwaukee	1,223.908
Boston	1,007.712	Cincinnati	1,225.763
Chicago	1,022.602	Memphis	1,231.066
Portland	1,037.523	Tampa	1,243.139
Salt Lake City	1,045.336	Jacksonville, Fla.	1,262.410
Seattle	1,048.947	Austin	1,268.820
Philadelphia	1,056.452	Rochester	1,297.772
New Orleans	1,061.193	Louisville	1,302.564
Washington, D.C.	1,072.714	St. Louis	1,303.700
San Diego	1,076.496	Detroit	1,308.164
West Palm Beach	1,083.576	Norfolk	1,313.868
Phoenix	1,089.006	Atlanta	1,317.805
Las Vegas	1,092.267	Charlotte	1,337.002
Cleveland	1,096.149	San Antonio	1,362.398
Providence	1,116.658	Orlando	1,367.412
Buffalo	1,118.312	Nashville	1,368.480
Los Angeles	1,128.605	Indianapolis	1,383.838
Columbus	1,144.620	Houston	1,407.333
Miami	1,146.274	Raleigh	1,460.261
Denver	1,157.805	Greensboro	1,528.339
Minneapolis	1,170.481	Grand Rapids	1,630.109
Dallas	1,183.744		

Source: Author's calculations based on Department of Transportation, "2001 National Household Travel Survey" (nhts.ornl.gov/2001 [May 2006]).

a. For purposes of this analysis, the average household consists of four people and has an annual income of $45,000.

These differences in gasoline consumption are due to more than the fact that suburbanites drive more than center city or near-center city residents. They also reflect the fact that suburbanites and urbanites often

high levels of local environmental quality and would have a small overall ecological footprint per capita. Sprawling cities have a larger ecological footprint than denser cities. As climate change takes place, will more sprawled cities suffer a reduction in their quality of life? The answer hinges on the physical location of such cities. If a sprawled Atlanta suffers more heat waves from climate change, then its local quality of life will decline both to due to more public health challenges caused by the summer heat and due to the disamenities of high humidity.

choose different vehicles. Based on data from the *2001 National Household Travel Survey*, 19.1 percent of suburban households own sports utility vehicles (SUVs), compared to 12.1 percent for urban households.[19] It is true that wealthier people live in the suburbs, but even controlling for household income, the probability of owning an SUV is much higher for a suburban household than for an observationally identical household based in the center city.

SPRAWL AND ROAD BUILDING. Increased vehicle use also has indirect environmental effects. For example, as more drivers drive more miles, demand for new roads grows. Road-building programs can have significant environmental costs. As a report from the National Resources Defense Council explains, "Sprawl also threatens water quality. Rain that runs off roads and parking lots carries pollutants that poison rivers, lakes, streams and the ocean. . . . As the impervious surfaces that characterize sprawling development—roads, parking lots, driveways and roofs—replace meadows and forests, rain no longer can seep into the ground to replenish our aquifers."[20]

The consequences of this land transformation include increased soil erosion and sharp increases in stream peak flow volumes, which increases flooding risk. In addition, sewer systems are often overwhelmed by the rapid runoff of storm water from roads.

SPRAWL AND PUBLIC TRANSIT. Equally important, sprawl reduces demand for public transit: use has been declining for more than forty years. In 1960, 22 percent of workers took public transportation or walked, and 64 percent traveled by car, either alone or in a carpool. By 1980 the share of those taking public transportation or walking had fallen by half: 6.4 percent took public transportation, 5.6 percent walked, and 84 percent drove. Since then, the proportion of workers taking public transportation has declined further, to 5.3 percent in 1990 and 4.7 percent in 2000.

---

19. This data set covers 11,201 households living in metropolitan areas. For the purposes of this analysis, urban households reside in census tracts with a population density greater than the sample median; suburban households live in census tracts with a population density below the sample median. The probability of owning an SUV in the entire sample is 15.2 percent. See U.S. Department of Transportation, "National Household Travel Survey—Downloads" (nhts.ornl.gov/2001/html_files/download_directory.shtml [May 2006]).

20. Otto and others (2003, p. 1).

Table 7-2. *Fraction of Workers Commuting by Public Transit, 1970–2000*[a]

Metropolitan statistical areas[b]	1970	1980	1990	2000
With rail transit in 1970				
Boston	0.18	0.14	0.14	0.15
Chicago	0.26	0.21	0.19	0.17
Cleveland	0.13	0.10	0.06	0.05
New York	0.45	0.37	0.37	0.38
Philadelphia	0.23	0.16	0.13	0.11
Pittsburgh	0.16	0.13	0.09	0.08
San Francisco	0.18	0.19	0.17	0.17
Total	0.30	0.25	0.23	0.23
With no transit in 1970 that constructed rail transit 1970–2000				
Atlanta	0.09	0.09	0.06	0.05
Baltimore	0.15	0.11	0.09	0.07
Buffalo	0.11	0.07	0.05	0.04
Dallas	0.06	0.04	0.03	0.02
Denver	0.05	0.06	0.05	0.05
Los Angeles	0.05	0.07	0.07	0.07
Miami	0.08	0.06	0.05	0.05
Portland	0.06	0.09	0.06	0.07
Sacramento	0.02	0.04	0.03	0.03
Salt Lake City	0.02	0.05	0.03	0.03
San Diego	0.04	0.04	0.04	0.04
San Jose	0.03	0.04	0.03	0.04
St Louis	0.09	0.06	0.03	0.03
Washington	0.17	0.16	0.16	0.14
Total	0.08	0.08	0.06	0.06
With no rail transit in 2000	0.05	0.03	0.02	0.02
All	0.12	0.08	0.07	0.06

Source: Baum-Snow and Kahn (2005).
a. For workers living within twenty-five miles of the central business district.
b. Defined as all tracts within twenty-five miles of the central business district.

Table 7-2 illustrates the large aggregate declines in the fraction of urban commuters using public transit. This dwindling usage occurred in "old transit" cities, such as New York City and San Francisco (metropolitan areas with historically high transit use and significant rail infrastructure in 1970); "new transit" cities, such as Dallas (metropolitan areas that established significant rail transit infrastructure after 1970); and "no transit" cities (metropolitan areas without rail transit in 2000). In old transit cities, the proportion of metropolitan-area workers commuting by public transit fell from 30 percent in 1970 to 23 percent in

1990. In cities where buses were the only public transit option, the fraction of public transit users dropped from 5 percent to 2 percent.[21]

Income growth plays some role in explaining this trend. As household incomes increase, people are less likely to use public transit, which is typically slower than commuting by car. Car travel takes about two minutes per mile for commutes under five miles. In contrast, bus commuting takes more than three minutes per mile for commutes under five miles. In addition, the average bus commuter waits nineteen minutes for a bus to arrive.[22] As a result, other than in a few major cities such as New York City, Boston, and San Francisco, richer people are unlikely to use public transit. Using data from the 2000 Census of Population and Housing for 41,301 urban census tracts, I estimate the following ordinary least squares regression:[23]

$$\text{Percent commute using public transit} = \\ \text{Metropolitan area fixed effects} - 4.26*\log(\text{income}) \\ + 3.10*\log(\text{population density}).$$

Controlling for metropolitan area fixed effects acknowledges the fact that metro areas differ with respect to their investments in public transit. For this sample of census tracts, the mean of the dependent variable is 8.5 percent. This regression highlights the fact that household income and sprawl both play key roles in determining public transit use. Holding a community's population density constant, I find that doubling household income reduces the share of workers who commute using public transit by 3 percentage points. Holding household income constant, doubling a community's population density (that is, urbanization) increases the share of workers who commute using public transit by 2.1 percentage points. Data from the 1990 census on individual cities support this point. In Boston 36 percent of workers who lived and worked in the center city commuted on public transit, compared to 5 percent of workers who lived and worked in the suburbs. In Chicago only 1.4 percent of workers who lived and worked in the suburbs commuted by public transit. In the year 2000, 10.2 percent of workers who lived within five miles of the central business district com-

---

21. Baum-Snow and Kahn (2005).
22. Glaeser, Kahn, and Rappaport (2000).
23. See Baum-Snow and Kahn (2005) for details about how the data set was constructed.

muted via public transit whereas only 5 percent of workers living more than five miles from the central business district used public transit for their commute.[24]

As households grow richer and move to the suburbs, public transit use rates are not the only casualty. Political support for train, subway, and bus systems also tends to decline, as a strong constituency in favor of "car-friendly" policies rises in its place. Higher-income individuals who commute by car find it in their interest to support low gasoline taxes and highway construction. By contrast, given that public transportation is used primarily by the poor, the "car class" has little stake in keeping it up. California's voters provide evidence supporting this claim. In 1994 they voted on Proposition 185, which would have imposed a 4 percent tax on gasoline.[25] Using data from the University of California at Berkeley's Statewide Database, I examine the share of voters who voted in favor of Proposition 185 as a function of how far they live from the central business district of a major metropolitan area.[26] Figure 7-1 examines voting patterns for Californians who live within twenty miles of a major metropolitan center. For each mile from the city center, I calculate the share of voters who voted in favor of the gasoline tax. The

---

24. This fact is generated using the data reported in Baum-Snow and Kahn (2005). The data set includes all census tracts within twenty-five miles of a central business district.

25. "This measure imposes a 4 percent sales tax on gasoline not diesel fuel beginning January 1, 1995. This new sales tax is in addition to the existing $.18 per gallon state tax on gasoline and diesel fuel and the average sales tax of approximately 8 percent imposed by the state and local governments on all goods, including gasoline. Revenues generated by the increased tax will be used to improve and operate passenger rail and mass transit bus services, and to make specific improvements to streets and highways. The measure also contains various provisions that generally place restrictions on the use of certain state and local revenues for transportation purposes. . . .

". . . Proponents include officials from the Congress of California Seniors, the Coalition for Clean Air, the Planning and Conservation League, Citizens for Reliable and Safe Highways, and the California Public Interest Research Group. . . .

". . . Opponents include officials from the California Transportation Commission, the California Highway Users Conference, the California Taxpayers' Association, the California Business Alliance, and the Alliance of California Taxpayers and Involved Voters." See Mary Beth Barber, "#185 Public Transportation Trust Funds. Gasoline Sales Tax," *California Journal* (www.calvoter.org/archive/94general/props/185.html [April 2006]).

26. The Institute of Governmental Studies, University of California at Berkeley, provides data on proposition voting counts for each California census tract. See "Statewide Database" (http://swdb.berkeley.edu [May 2006]).

Figure 7-1. *Voter Support for Proposition 185 as a Function of Residential Distance from a Central Business District, California, 1994*

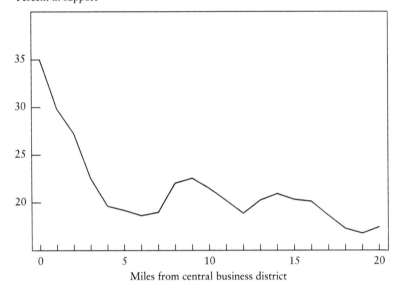

Source: Author's calculations based on data from University of California at Berkeley, "Statewide Database" (http://swdb.berkeley.edu [May 2006]).

figure shows that people who live within one mile of the city center are 15 percentage points more likely, on average, to vote in favor of higher gasoline taxes than people who live more than five miles from the city center.

Despite this dynamic, cities continue to invest in public transit—at least in part because the federal government subsidizes such investments at rates as high as 75 percent. Over the last thirty years, Atlanta, Baltimore, Boston, Chicago, Dallas, Denver, Los Angeles, Miami, Portland, Sacramento, San Diego, San Francisco, San Jose, St. Louis, and Washington, D.C., have all spent large sums of money (as high as $300 million per mile in Los Angeles) to build or upgrade their light-rail systems. Urban economists have been quite pessimistic about whether these irreversible investments will pay off in the form of significant increases in ridership and reduced vehicle use, and so far their pessimism has been borne out.[27] As reported in table 7-2, across all fourteen cities that

27. See, for example, Kain (1990, 1997, 1999).

invested in rail transit, the share of workers commuting by public transit declined from 8 percent in 1970 to 6 percent in 2000.

Rail transit construction is particularly unlikely to pay off in sprawling cities because most transit systems are geared toward serving central business districts.[28] If a significant share of employment has already moved to the suburbs, people are highly unlikely to use a commuting option that takes them downtown. Combine employment suburbanization with rising household income, and the result is that relatively few drivers abandon their cars to ride the train. Instead, the main beneficiaries of new train lines tend to be bus commuters, who switch from slow buses to faster trains.

However, the picture is not entirely bleak. In two cities—Boston and Washington, D.C.—public transit use has been on the rise.[29] The Washington Metropolitan Area Transit Authority reports that ridership between 1994 and 2004 grew by 27 percent.[30] These gains suggest that in some cases, improvements in the quality of rail transit can attract noncommuters, such as seniors, tourists, and teenagers, as well as suburbanites who work downtown.

Land Consumption

Another important consequence of sprawl is an increase in per capita land consumption. Based on the 1999 American Housing Survey, subur-

---

28. Baum-Snow and Kahn (2005).
29. Baum-Snow and Kahn (2000, 2005). Using data from the U.S. decennial Census of Population and Housing from 1970 through 2000, Nathaniel Baum-Snow and I identify census tracts (geographic blocks of roughly 4,000 people) that were more than two kilometers away from the nearest rail transit line but due to rail transit expansion and construction are subsequently within two kilometers from the nearest rail transit line. We call these tracts the "treated tracts" and compare changes in public transit use over time in these tracts relative to similar "control tracts" within the same metropolitan area that have remained far from rail transit lines. Washington, D.C., is the metropolitan area where we find the largest increases in ridership. Washington's 1970s expansion increased public transit use by 9.2 percentage points in tracts more than ten kilometers from the central business district (CBD). Its 1980s expansion and 1990s expansions increased use in treated tracts more than ten kilometers from the CBD by 6.6 and 3 percentage points, respectively. For residential communities located within ten kilometers of Washington's CBD that received increased access to rail transit, we cannot reject the hypothesis that there has been no increase in the share of workers commuting using public transit.
30. See *Washington Post*, "Metro Usage" (www.washingtonpost.com/wp-srv/metro/daily/graphics/metrousage_091004.html {May 2006]).

ban homeowners have 35 percent more exterior space and 6 percent more interior space than center city homeowners in the same metropolitan area and with the same income.[31] As a result suburbanization has been associated with a significant decline in open space across the country. Between 1982 and 1992, urban and suburban growth caused the amount of farmland in the United States to fall by four million acres—an area two-thirds the size of Vermont.[32]

While measuring the amount of land that has been suburbanized is a straightforward task, determining the costs of this transformation poses much greater challenges.[33] For example, one ecological study concluded that "every 1 per cent increase in watershed area covered by wetlands decreased flood peaks in streams by nearly 4 percent."[34] Thus, if a new suburban development project reduces the size of a wetland area by 5 percent, flood peaks in streams are likely to increase by 20 percent. How costly is this impact? To answer this question, detailed information is needed on how many people live near the streams and thus are at risk of being exposed to flooding. In addition, one would need to know whether their homes are sturdy enough to withstand the flood and whether the residents have access to radios and other equipment that would help them evacuate in the event of a major flood. To an economist the challenge here is to determine the value of the ecological services that this land was providing before suburbanization took place. If a value cannot be imputed to the lost natural capital, then one really cannot know how costly suburban development has been.[35]

---

31. This data set includes information on 11,691 homeowners. See HUD User, "Data Sets—American Housing Survey" (www.huduser.org/Datasets/ahs.html [May 2006]).

32. See Population Connection, "Population and the Environment" (www.populationconnection.org/Communications/ED2002WEB/PopEnvFactSheet2002.pdf [May 2006]).

33. These costs will differ across metropolitan areas depending on their geography and nearby natural reources. For example, an interdisciplinary research team has been investigating how growth in Maryland's Montgomery County will affect local watersheds. The ecological cost of the anticipated population growth hinges on where this growth occurs. Conservation easements in large undeveloped areas within this county could play a major role in protecting natural capital. See Palmer and others (2002).

34. McCully (1996, p. 191).

35. This point echoes a theme raised in chapter 2, which discussed an explicit formula for a green city index. This equation highlighted that a city's "greenness" hinges on its objective levels of environmental quality (that is, such indicators as air

This problem is particularly acute when it comes to forms of natural capital, such as species diversity, for which no markets exist. Ideally, land conversion would take place primarily in areas where the risk of species extinction is low. But once again, externalities get in the way. The quality of life in Atlanta, for example, is unlikely to be threatened by the decline of the Cherokee darter, one of at least seven imperiled species in Bartow County.[36] Consequently, local politicians have little incentive to protect the darter by slowing development. Fortunately, the growth of open space initiatives throughout the United States provides one implicit insurance policy against the environmental impacts of sprawl. Between 1998 and 2003, for example, New Jersey residents voted on 237 different referenda that sought to provide incentives for local governments to raise taxes for open space preservation.[37] If ecologists have a voice in helping politicians decide where such preservation should take place, suburban land consumption will impose fewer ecological costs.

Water Consumption

While all major U.S. cities are sprawling, the greatest growth has taken place in warmer, arid areas, such as Las Vegas and Phoenix. One result has been a sharp increase in water demand relative to supply. Yet water prices rarely reflect this growing scarcity. A recent study of prices at 1,980 water supply systems across the United States found no evidence that an area's climate was an important factor in determining water prices in the early 1990s.[38] Intuitively, this means that residents of arid cities do not pay a price premium for the water they use. Consequently, they have little incentive to economize, and the ecological footprint of these cities continues to grow.

Growing urban water demand can also have other effects. For example, farmers may become increasingly likely to sell their water rights to

---

and water pollution). In addition, this index requires prioritizing the relative importance of these different measures of urban environmental quality. Environmental economists continue to try to measure how much people value improvements in nonmarket environmental quality. Chapter 2 provided a brief overview of what researchers have learned about environmental valuation using hedonic real estate methods.

36. Ewing and others (2005).
37. Kotchen and Powers (forthcoming).
38. See Kirshen and others (2004).

municipalities.³⁹ Is this a bad thing? Not necessarily, since farmers can always choose *not* to sell their water rights. Rational farmers will compare their profits from continuing to grow crops using their water allotments to what they could earn if they sold some of their water to urbanites in arid areas.

To help match water demand to supply in fast-growing, arid metropolitan areas, some state and local governments have tied the approval of residential development permits to the availability of adequate water supply.⁴⁰ Other politicians have turned to technological mandates. In the 1980s, for example, severe droughts prompted many California cities to install water meters and require residents and firms to participate in conservation efforts. In some cities residents were banned from watering their lawns more than once a week and required to install low-flow showerheads in their homes. According to studies, by installing such devices, the typical home could cut its average daily water use from seventy-four to fifty-two gallons.⁴¹

But encouraging the diffusion of these green technologies is not sufficient to shrink a city's per capita footprint. Consumers must also face price incentives. As Christopher Timmins observes,

> With all of this technical potential, however, low-flow toilets and showerheads have proven most effective when adopted in conjunction with increasing prices for water at the margin. . . . Only with such prices would we expect consumers to avoid illegal alterations designed to make the low-flow devices function like traditional fixtures. This has clearly been the case in Los Angeles, where the adoption of increasing block rate price schedules (e.g., a household's marginal price triples if usage exceeds 13,000 gallons a month) in conjunction with the subsidized adoption of low-flow toilets and showerheads resulted in a 12 percent reduction in water use between 1989 and 1995, despite a growing urban population.⁴²

In addition to more efficient technologies and higher prices, more compact urban development could reduce water demand. California planners claim that increasing residential density from four units per acre to

---

39. Libecap (2005).
40. Hanak and Chen (forthcoming).
41. Timmins (2002).
42. Timmins (2002).

five units per acre could reduce water usage by roughly 10 percent.[43] This estimate provides a sense of how much sprawl contributes to increased water consumption.

Fighting Sprawl with Smart Growth

Some cities have sought to fight sprawl and its environmental consequences by implementing "smart growth" policies. Kent Portney has summarized these efforts by constructing green policy indicators covering thirty areas.[44] These indicators measure, for example, whether cities redevelop brownfields, use zoning to delineate environmentally sensitive growth areas, provide tax incentives for environmentally friendly transportation, limit downtown parking spaces, and purchase or lease alternatively fueled vehicles.[45] Based on these indicators, Portney calculates a sustainability score for twenty-four cities. For example, if a city is active in seventeen of the thirty categories, its score is 17. Using this approach, Portney identifies the seven most sustainable cities as Seattle, Scottsdale, San Jose, Boulder, Santa Monica, Portland, and San Francisco. Each of these cities received a score of 23 or higher.

How do these "green policy cities" differ from cities that have made fewer investments in smart growth? Based on 1990 census data, table 7-3 reports population-weighted means in several demographic categories for all 1,083 U.S. cities with more than 25,000 people and for the seven top-scoring cities and the remaining "brown policy cities" in Port-

---

43. Department of Water Resources, "Chapter 20: Urban Land Use Management," *California Water Plan Update 2005. Volume 2: Resource Management Strategies* (www.waterplan.water.ca.gov/docs/cwpu2005/Vol_2/V2PRD20_urbland.pdf [April 2006]).

44. Portney (2003).

45. The other indicators include ecoindustrial park development, cluster or targeted economic development, ecovillage project or program, land use planning programs, comprehensive land use plan that includes environmental values, operation of inner-city public transit, car pool lanes, bicycle ridership program, pollution prevention, household solid waste recycling, industrial recycling, hazardous waste recycling, air pollution reduction program, recycled product purchasing by city government, Superfund site remediation, asbestos abatement program, lead paint abatement program, energy and resource conservation, green building program, renewable energy use by city government, energy conservation effort, alternative energy offered to consumers, water conservation program, a single government agency responsible for implementing sustainability as part of a citywide comprehensive plan, and involvement of the city council, mayor, business community, and the general public (Portney 2003).

Table 7-3. *Characteristics of Smart Growth Cities versus Other U.S. Cities, 1990*[a]
Units as indicated

Demographic characteristics	Cities ranked by Portney		
	All cities[b]	Smart growth cities[c]	All other cities[d]
Population density (people per square mile)	7,501.478	7,541.242	3,965.614
Black (percent)	19.00	7.51	18.44
White (percent)	68.54	69.27	72.41
Hispanic (percent)	14.98	13.18	13.50
Median age (years)	32.113	33.912	30.954
Foreign born (percent)	13.61	21.44	8.48
College graduates (percent)	22.48	32.31	22.65
Median household income (1990 dollars)	29,895.61	35,265.87	26,757.11
Below poverty line (percent)	15.96	11.73	16.99
Commuting via public transit (percent)	12.84	14.88	7.77
Average travel time (minutes)	23.55	23.92	20.97

Source: Author's calculations based on Portney (2003) and on 1990 census data from University of Virginia, Geospatial and Statistical Data Center, "1994 City Files" (fisher.lib.virginia.edu/collections/stats/ccdb/city94.html [May 2006]).
a. Values shown are population-weighted means.
b. Values for 1,083 cities with a population greater than 25,000.
c. Boulder, Portland, San Francisco, San Jose, Santa Monica, Scottsdale, and Seattle.
d. Austin, Boston, Brownsville, Cambridge, Chattanooga, Indianapolis, Jacksonville, Milwaukee, New Haven, Olympia, Orlando, Phoenix, Santa Barbara, Tampa, and Tucson.

ney's data set. This table reveals stark differences between Portney's two sets of cities. The cities that enact green policies are wealthier and more educated. In green policy cities, 32.3 percent of adults have college degrees whereas only 22.6 percent of adults have college degrees in their brown policy counterparts. The average poverty rate is 11.7 percent in the green policy cities and 17 percent in the brown policy cities. Average population density is 90 percent higher in green policy cities than brown policy cities, and the share of workers who commute by public transit is 14.9 percent in the first group versus 7.8 percent in the second. These large differentials support the hypothesis that richer governments are more likely to enact greener policies. They also support the hypothesis that education plays an important role in stimulating environmentalism.

While the results in table 7-3 document the demographic differences between cities that do and do not implement smart growth policies, the results provide no evidence on whether green policy cities are better places to live. In addition to their environmental benefits, green policies can have potentially undesirable consequences. For example, Portland's urban growth boundary mitigates sprawl by promoting infill development.[46] By limiting the supply of new lots that can be developed, it also helps raise the price of existing homes.[47] As a result homeowners win twice: the city becomes greener, and the value of one of their most important assets goes up. But renters are likely to see their cost of living rise, and lower-income groups, including minorities, will find it increasingly difficult to purchase homes.[48]

Local growth control policies can also have the unintended effect of encouraging sprawl. Zoning policies in one community can deflect growth toward the suburban fringe.[49] Consider Marin County north of the Golden Gate Bridge in San Francisco. As William Fischel observes,

> It has large amounts of open space on which development could easily occur but does not. Tens of thousands of commuters from far away suburbs and exurbs pass through the Marin County corridor on U.S. Route 101 on their way to work in San Francisco. Marin's open space is an asset for those who live near it and it probably provides some pleasures for those who drive through it daily. But it also represents an enormous waste in the form of excessive commuting and displacement of economic activities to less productive areas.[50]

This example points to an interesting urban tradeoff. The voters in Marin County have chosen to have higher taxes in return for preserving open space. This "greens" the county and raises the environmental aesthetics for the people who live there. But as Fischel points out, an unintended consequence of such an open space policy is that new homeowners are pushed even further out into the fringe in pursuit of affordable housing. If these households commute to downtown San Francisco to work and shop, the ecological footprint of such sprawled suburbanites is much larger than it would have been had Marin County not preserved

---

46. Phillips and Goodstein (2000); O'Toole (2001).
47. Katz and Rosen (1987), Glaeser, Gyourko, and Saks (2005)
48. Collins and Margo (2003); Kahn (2001a).
49. Another important explanation for suburban growth in farmland areas is that such areas have not been built upon. Since housing is durable, it is much cheaper for a developer to purchase a large open field and transform it into multiple large houses rather than to purchase hundreds of small suburban homes built in the 1950s in an inner suburban ring, knock them down, and start over building something fancier in their place.
50. Fischel (1999, p. 162).

so much land as open space. In this sense Marin County's open space policy makes the Bay area less, not more, green.

Conclusion

Millions of people in the United States are moving to low-density cities with warmer climates that offer larger, newer homes and a more pleasant outdoor environment. Within older metropolitan areas such as Chicago and Detroit, the center city population is shrinking or barely growing while growth in population and employment is occurring at the urban fringe.

By some measures sprawl has not had a negative impact on urban sustainability. For example, there is little evidence of air quality degradation in growing U.S. cities, as discussed in chapters 5 and 6. But environmentalists would argue that sprawl is causing the ecological footprint of the fastest growing cities to increase more rapidly than it otherwise would. Suburbanites, for example, consume more gasoline, which increases greenhouse gas production and increases the probability of climate change. This suggests that the major environmental costs of sprawl may be global rather than local. I will return to this point in the next chapter.

There is a certain irony in the fact that environmentalists focus so much attention today on the costs of sprawl. In the past many of the country's urban environmental problems revolved around high-density living. Before transportation innovations such as the internal combustion engine, rising urban population levels translated into higher residential densities. For example, the population of lower Manhattan increased by 74 percent between 1820 and 1850.[51] As Martin Melosi writes,

> Such crowded conditions provided fertile ground for health and sanitation problems. Many workers had little choice but to live in the least desirable sections of the city, usually close to smoky factories or near marshy bogs and stagnant pools. City services, especially sewage and refuse collection, failed to keep up with demand. Smoke from wood burning and coal burning stoves and fireplaces fouled the air, and the noise level reached a roar.[52]

51. Melosi (2001, p. 31).
52. Melosi (2001, p. 35).

Sprawl has made these problems a distant memory.

But less desirably, it may also threaten to make the collective efforts that helped solve these problems a relic of the past. When rich and poor clustered together in center cities, wealthy urbanites could not so easily escape the problems of their less fortunate neighbors. Pollution or disease spread easily and quickly from the tenements of the poor to the mansions of the rich. As a result upper-bracket taxpayers were more likely to support policies that improved the living conditions of the worst off.[53] For example, as Werner Troesken points out, "In a world where blacks and whites lived in close proximity 'sewers for everyone' was an aesthetically sound strategy. Failing to install water and sewer mains in black neighborhoods increased the risk of diseases spreading from black neighborhoods to white ones."[54]

Today, suburbanization has greatly increased the physical distance between the middle and upper middle class and the poor. Consequently, it is much easier and less risky for wealthier taxpayers to ignore the problems of those who are less well off. This is bad news, not just for poorer urbanites but for the center cities that suburbanites have left behind.

---

53. Troesken (2004).

54. Troesken (2004, p. 10). Chinatown in San Francisco offers an interesting case study (Craddock 2000). Within the city typhoid rates were highest in the immigrant Chinatown area. To reduce the prospects of a public health crisis emerging from there, proactive steps were taken to invest in public health. Civic leaders recognized that this community interacted with the native community and hence there existed the possibility of disease contagion. A rather large percentage of Chinese immigrants who lived in Chinatown worked outside of Chinatown in laundries, as cooks, and as domestic workers. Many also traveled to outlying farm areas, transporting produce and other commodities from truck farms to the city of San Francisco.

CHAPTER 8

Achieving Urban and Global Sustainability

Soon most people around the world will live and work in capitalist cities. Thus the quality of life of billions of people hinges on whether free-market economic development fosters the growth of green cities. This issue has generated a spirited debate. On one side stand environmentalists, who use the ecological footprint as their key indicator of overall sustainability. They argue that development and urbanization translate into greater demand for resource-intensive goods such as cars and free-standing homes. Producing, using, and disposing of such goods will require constantly increasing resources and generate growing amounts of pollution and waste. The result, environmentalists fear, will be both browner cities and a browner world.[1]

On the other side of the debate stand many economists, who believe that economic growth will cause the quality of life of billions of urbanites to improve. As evidence they point to long-term increases in life expectancy and human height, which suggest that development has brought about considerable gains in the battle against pollution and disease.[2] In addition, they offer the environmental Kuznets curve (EKC) as an explanation for how such gains can come about. According to the EKC, economic growth initially increases pollution, but once a certain

---

1. See Diamond (2005).
2. Fogel (2000).

income level has been reached, further development is associated with environmental progress.

To the left of the EKC turning point, the primary effect of income growth is to increase the scale of consumption and production. In other words, more cars clog city streets, smokestacks rise, the air becomes dark with smog, and water and sanitation services experience growing strain. However, once the turning point is reached, consumer preferences, new technologies, and regulations more than offset these effects. Richer consumers are more likely to demand higher-quality products, and this quality effect can sometimes mitigate environmental problems, even if that is not the consumers' primary goal. Richer and more educated workers will also tend to work in cleaner industries and settle in cities where such industries have replaced the smokestack production of decades past. Finally, around the world there is evidence that regulation becomes more stringent as national incomes increase. As proenvironment incentives and policies are implemented, the ecological footprint of growing cities may stabilize or even decrease.

The EKC is a powerful idea, and evidence on a variety of urban environmental indicators, such as air and noise pollution, tends to support this hypothesis. The EKC is also more subtle than some of its critics (and even some proponents) believe. It does not assume that growth will automatically take care of environmental problems. Whether growth mitigates urban environmental problems hinges on the incentives for urban consumers, producers, and politicians. This book has paid careful attention to each of these key decisionmakers to understand why their actions sometimes "brown" cities and in other cases help to "green" them.

But the EKC has limitations. By focusing on the relationship between per capita income and environmental quality, it offers an incomplete picture of the consequences of urban growth. In fact, urban growth incorporates three trends: income growth, population growth, and spatial growth or sprawl. These trends are closely related. Rising urban incomes, for example, trigger migration to cities. The resulting population growth can degrade urban environmental quality by increasing the scale of consumption and production. In addition, as income growth continues, urban populations tend to sprawl. Richer urbanites often want more space and newer housing, which is easier to find in low-density, vehicle-friendly suburbs. Suburbs also offer more green amenities, such as open space and cleaner air. But by moving out of the center city, new suburbanites contribute to broader environmental problems.

As people suburbanize, more land is paved over, water and gasoline consumption increases, and support for public transit falls.

Does this mean that the environmentalists are right? Should economists abandon their belief in the underlying optimism of the EKC? There is no simple answer to these questions, as this book has shown. In many important cases, such as air quality, this optimism seems justified. But even if the EKC proves to be an accurate description of long-term trends, for cities and countries on the upward-sloping portion of the curve, the environmentalists' concerns may have much greater relevance here and now. Imagine urban growth as a race among its three component trends. If income grows far more rapidly than population or sprawl, then urbanites' environmental quality of life may rapidly improve. But if income growth is slow while population growth and sprawl sprint ahead, the outlook is bleak.

This problem is exacerbated when the EKC turning point lies far to the right. For localized problems, such as water and air pollution, the turning point is likely to occur at lower levels of per capita income—although note that most of the world's countries have yet to achieve per capita income in the range of $6,000 to $8,000, which is where most studies indicate EKC turning points lie. When pollution is local, urbanites directly experience the health costs from allowing a pollution problem to fester. This motivates them to support government regulation. But when environmental problems involve large externalities, the turning point (given current technologies) may lie far out of reach, at least for the generations alive today. Greenhouse gases, which have been implicated as a major contributor to climate change, are the leading example of this problem. The release of such gases poses a distinct "tragedy of the commons" problem: since the atmosphere is common property, no one nation has an incentive to unilaterally reduce its emissions.

Climate Change and the Future of Cities

The International Panel on Climate Change, an offshoot of the World Meteorological Organization and the United Nations Environment Programme, has predicted that average temperatures worldwide will increase by 2.5 degrees Celsius (4.5 degrees Fahrenheit) by 2100.[3] Low-

---

3. See Union of Concerned Scientists, "The Intergovernmental Panel on Climate Change" (www.ucsusa.org/global_warming/science/the-intergovernmental-panel-on-climate-change.html [May 2006]).

lying and coastal cities will be particularly hard hit by this trend, due to anticipated changes in sea levels as polar ice thaws.

The Mayor of London's office recently catalogued the ways in which climate change is likely to affect the city. Some of these effects have already been felt. As a planning document points out, "Sea-level rise relative to the land is now widely accepted as occurring at 6mm/year at high tide in the London area."[4] Tidal flooding has therefore become increasingly likely, especially in east London, where much future development is slated to take place. In addition, temperatures are likely to rise, changing the demand for winter heating and summer cooling, as well as seasonal mortality patterns; river flows are likely to rise in winter and dip in the summer, potentially compromising water quality; warm summer droughts may become more common, threatening wetlands and possibly contributing to the spread of disease; and overall rainfall may increase by as much as 10 percent, along with the unpredictability of the weather. The economic impact of these changes will be felt in a wide range of sectors, including construction, transport, finance (particularly the insurance industry), and tourism.

Rich cities, such as London and New York, have the greatest resources to cope with climate change, but the challenges they face will remain significant, especially for the least well off. For example, climate change models predict that by the 2050s, summer temperatures will have risen by 2.12–2.75 degrees Celsius.[5] The number of annual heat wave days could rise from an average of fourteen days (for the period from 1900 to 1997) to roughly fifty. (By way of comparison, the 1995 Chicago heat wave, which claimed roughly 500 lives, lasted only five days.) Poor urbanites, who are already hardest hit by extreme weather, will bear the brunt of this impact.[6]

Cities in poorer countries will not only have fewer resources for coping with these problems, they may also face additional challenges as a result of climate change. As growing conditions change, many farmers in poorer countries will find it difficult to adapt. As poverty increases among farming families in largely rural nations at low latitudes,

---

4. Mayor of London, "The London Plan" (www.london.gov.uk/approot/mayor/strategies/sds/london_plan_download.jsp [April 2006]).

5. Climate Change Information Resources, "What Changes in Climate Are Projected for the Region?" (ccir.ciesin.columbia.edu/nyc/ccir-ny_q2a.html [April 2006]).

6. The disadvantaged often bear the brunt of the consequences caused by environmental challenges. In the 1995 Chicago heat wave, elderly and black residents were overrepresented among the people who died (Klinenberg 2002).

migration to the cities will increase, straining urban resources and infrastructure.

In the long run, urban growth may help reduce the likelihood of climate change. Urbanization reduces national population growth by providing greater economic opportunities for women outside the household and eliminating the need for children to do farm work. Both trends tend to depress birth rates, which can ultimately reduce a nation's ecological footprint. In the short to medium term, however, urbanization is likely to exacerbate global warming. Researchers have found that more urbanized nations produce more greenhouse gas emissions as a result of higher living standards and higher demand for transportation, energy, water, and other resources and services.[7] And although there is some evidence that emissions of greenhouse gases, such as carbon dioxide, follow the EKC, the turning point lies very far to the right—around $20,000 in per capita income.[8] Most of the world's economies lie to the left of this point. Consequently, as these economies grow, greenhouse gas production will increase.

Can advances in technology break the link between economic activity and the probability of climate change? In the past, technology has often helped solve the problems caused by population growth and increases in resource use, as William Nordhaus explains,

> For most of the nineteenth and twentieth centuries, concerns about resource exhaustion have receded as technological change has outpaced the modest degree of resource exhaustion. New seeds and chemical fertilizers have more than offset the need to move cultivation to marginal lands; advances in finding and drilling for oil have countered the need to drill deeper and in harsher climates; and modest pollution-abatement investments have allowed economic growth to continue while lowering concentration of many toxic substances. In short, for the past two centuries, technology has been the clear victor in the race with depletion and diminishing returns.[9]

But if technology is to come to the rescue, economic actors—including countries, firms, and individuals—must have sufficiently strong incentives to reduce carbon dioxide and other greenhouse gas emissions.

---

7. Parikh and Shukla (1995).
8. Schmalensee, Stoker, and Judson (1998).
9. Nordhaus (1992, p. 38).

Economists have suggested that a cap and trade system would provide such incentives at the national level. Under this proposal each country would receive an allocation of rights to create carbon dioxide. If a nation exceeded its allocated quota, it would have to purchase pollution permits from another nation. This approach would encourage the adoption of greener techniques that could sharply reduce carbon dioxide emissions per dollar of gross national product. But major implementation challenges include the task of monitoring carbon dioxide emissions around the world and credibly committing to punish polluters who pollute without permits.[10]

In the absence of such market incentives, there is little evidence of significant reductions in carbon dioxide production, even in rich nations. Consider figure 8-1, which graphs per capita carbon dioxide emissions for the United States and Canada from 1960 to 2000. Between 1960 and 2000, emissions of carbon dioxide increased by 2.1 percent per year in the United States, but emissions per dollar of GDP declined by 1.8 percent per year. The picture is slightly brighter in Canada, where per capita emissions have generally declined since 1990.

While environmentalists argue that the United States should take the lead in scaling back its emissions, both the Clinton and Bush administrations have been slow to take action. Why? U.S. taxpayers compare the short-term costs of preempting climate change with what they perceive to be the benefits they will eventually enjoy. And relatively few conclude that reducing the risk of global warming in the future is worth higher prices for gasoline.[11]

Why are taxpayers ignoring the advice of professional ecologists?[12] One explanation may be blissful ignorance. Americans do not see nor

---

10. In the absence of a price incentive, growing population and rising income will translate into more greenhouse gas production. The tragedy of the commons problem will be exacerbated. Fortunately, the recent experience in the United States with the sulfur dioxide trading market and improvements in information technology offer the possibility that a market trading system could be feasible.

11. One surprising recent policy initiative is that seven northeastern states are embarking on a plan to reduce power plant emissions of carbon dioxide by 10 percent below current emissions levels by the year 2019. See "Regional Greenhouse Gas Initiative" (www.rggi.org [May 2006]). While this regional initiative will only have a negligible impact on reducing greenhouse gases, it will provide an opportunity to see how large are the costs of complying with carbon dioxide caps.

12. Some might point out that ecologists' past incorrect predictions have lowered their credibility with the public.

Figure 8-1. *U.S. and Canadian Carbon Dioxide Production, 1960–2000*

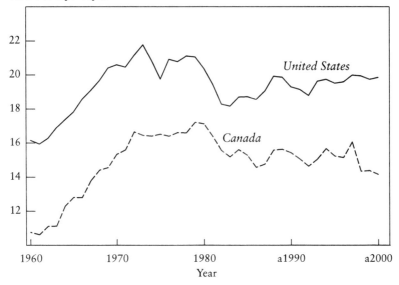

Source: Based on country profiles from World Bank, "World Development Indicators 2006—Key Development Data and Statistics" (www.worldbank.org [May 2006]).

want to acknowledge the possibility of an upcoming train wreck. A second explanation is that people in the United States are technological optimists who believe that the country has the resources to address this problem "when we get there."[13] A third explanation is that U.S. taxpayers are aware that much of the cost of climate change will be borne by strangers in other low-lying nations in the distant future. Climate change will raise quality of life in some cities and countries and lower quality of life in others, and predicting the differential effects of climate

13. Even if voters can be convinced that climate change is a real future threat, will they being willing to sacrifice today to preempt an event fifty years from now? An irony arises here: the typical U.S. voter might say to himself, "I can handle climate change on my own. I'll buy a better air conditioner or perhaps a coastal summer home." By increasing his own resource consumption and carbon dioxide production, this voter has insulated himself from climate change's consequences! If voters believe that they can cope with climate change, then this reduces the likelihood that they will elect politicians who push a proactive agenda for addressing this issue.

change remains a speculative exercise.[14] Nonetheless, as William Nordhaus and Joseph Boyer observe,

> The United States appears to be less vulnerable to climate change than many countries. This is the result of its relatively temperate climate, small dependence of its economy on climate, the positive amenity value of a warmer climate in many parts of the U.S., its advanced health care system and low vulnerability to catastrophic climate change . . . the economic impact of gradual climate change (that is, omitting catastrophic outcomes) is close to zero for a moderate global warming.[15]

Consequently, the overall trend for the future environmental health of U.S. cities is quite positive. Urbanites increasingly value quality of life and recognize that environmental quality is a key component of daily quality of life. Major cities will continue to invest in improving urban air and water quality and cleaning up and in restoring areas affected by past urban blight. In short, for richer cities the outlook is good. However, cities that are still climbing the environmental Kuznets curve slope may face decades of growing pollution ahead. In addition, they will be forced to struggle with the global environmental consequences of increased urban economic activity in both the developed and developing worlds.

---

14. I do not mean to imply that climate change represents a zero-sum game for U.S. cities. Losing warmer coastal cities (such as New Orleans) may lose more than winning cities (such as Minneapolis) gain. Economists have emphasized that if climate change is a gradual process, then the population will have time to adjust and this will reduce the social costs of this trend. For example, people who are sensitive to excess heat would migrate to more northern cities. Under this scenario, property owners in cities whose climate has become unpleasant would bear the major costs of climate change as their real estate's worth would decline.

15. Nordhaus and Boyer (2003, pp. 96–97).

References

Ades, Alberto, and Edward L. Glaeser. 1995. "Trade and Circuses: Explaining Urban Giants." *Quarterly Journal of Economics* 110, no. 1: 195–227.

Alberini, Anna, and others. 1999. "Valuing Mortality Reductions in India: A Study of Compensating-Wage Differentials." Working Paper 2078. Washington: World Bank.

Alesina, Alberto, Reza Baqir, and William Easterly. 1999. "Public Goods and Ethnic Divisions." *Quarterly Journal of Economics* 114, no. 4: 1243–84.

Alesina, Alberto, and Edward Glaeser. 2004. *Fighting Poverty in the U.S and Europe: A World of Difference*. Oxford University Press.

Alesina, Alberto, and others. 2003. "Fractionalization." *Journal of Economic Growth* 8, no. 2: 155–94.

Altshuler, Alan, and others, eds. 1999. *Governance and Opportunity in Metropolitan America*. Washington: National Academies Press.

Anand, Sudhir, and Amartya K. Sen. 1994. "Human Development Index: Methodology and Measurement." Human Development Report Office Occasional Paper 12. New York: United Nations Development Programme.

———. 2000. "The Income Component of the Human Development Index." *Journal of Human Development* 1, no. 1: 83–106.

Anderson, Terry, and Donald Leal. 2001. *Free Market Environmentalism*. New York: Palgrave Macmillan.

Arrow, Kenneth, and others. 2004. "Are We Consuming Too Much?" *Journal of Economic Perspectives* 18, no. 3: 147–72.

Arzaghi, Mohammad, and Vernon Henderson. 2005. "Why Countries Are Fiscally Decentralizing." *Journal of Public Economics* 89, no. 7: 1157–89.

Ashenfelter, Orley, and Michael Greenstone. 2004. "Using Mandated Speed Limits to Measure the Value of a Statistical Life." *Journal of Political Economy* 112, no. 1, part 2: S226–67.

Athey, Susan, and Scott Stern. 2002. "The Impact of Information Technology on Emergency Health Care Outcomes." *RAND Journal of Economics* 33, no. 3: 399–432.

Baden, Brett, and Don Coursey. 2002. "The Locality of Waste Sites within the City of Chicago: A Demographic, Social, and Economic Analysis." *Resource and Energy Economics* 24, no. 1–2: 53–93.

Bajari, Patrick and Matthew E. Kahn. 2005. "Estimating Housing Demand with an Application to Explaining Racial Segregation in Cities." *Journal of Business and Economic Statistic*s 23, no.1 (January): 20–33.

Bartelsman, Eric J., and Wayne Gray. 1996. "The NBER Manufacturing Productivity Database." Technical Working Paper 205. Cambridge, Mass.: National Bureau of Economic Research (NBER) (October).

Baum-Snow, Nathaniel. 2005. "The Effects of Changes in the Transportation Infrastructure on Suburbanization: Evidence from the Construction of the Interstate Highway System." Ph.D. dissertation, University of Chicago.

Baum-Snow, Nathaniel, and Matthew E. Kahn. 2000. "The Effects of New Public Projects to Expand Urban Rail Transit." *Journal of Public Economics* 77, no. 2: 241–63.

———. 2005. "Effects of Urban Rail Transit Expansions: Evidence from Sixteen Cities, 1970–2000." In *Brookings-Wharton Papers on Urban Affairs 2005,* edited by Gary Burtless and Janet Rothenburg Pack, pp. 147–206. Brookings.

Becker, Gary, and Casey Mulligan. 1997. "The Endogenous Determination of Time Preference." *Quarterly Journal of Economics* 112, no. 3: 729–58.

Becker, Randy, and Vernon Henderson. 2000. "Effects of Air Quality Regulations on Polluting Industries." *Journal of Political Economy* 108, no. 2: 379–421.

Beede, David, and David Bloom. 1995. "The Economics of Municipal Solid Waste." *World Bank Research Observer* 10, no. 2: 113–50.

Bento, Antonio M., and others. 2005. "The Impact of Urban Spatial Structure on Travel Demand in the United States." *Review of Economics and Statistics* 87, no. 3: 466–78.

Berger, Mark C., Glenn Blomquist, and Klara Sabirianova. 2003. "Compensating Differentials in Emerging Labor and Housing Markets: Estimates of Quality of Life in Russian Cities." Discussion Paper 900. Bonn, Germany: Institut zur Zukunft der Arbeit (IZA).

Berman, Eli, and L. Bui. 2001. "Environmental Regulation and Labor Demand: Evidence from the South Coast Air Basin." *Journal of Public Economics* 79, no. 2: 265–95.

Bhagwati, Jagdish. 2004. *In Defense of Globalization.* Oxford University Press.

Bin, Okmyung, and Stephen Polasky. 2004. "Effects of Flood Hazards on Property Values: Evidence Before and After Hurricane Floyd." *Land Economics* 80, no. 4: 490–500.

Bresnahan, Brian, Mark Dickie, and Shelby Gerking. 1997. "Averting Behavior and Urban Air Pollution." *Land Economics* 73, no. 3: 340–57.

Bresnahan, Timothy, and Dennis Yao. 1985. "The Nonpecuniary Costs of Automobile Emissions Standards." *RAND Journal of Economics* 16, no. 4: 437–55.

Burtraw, Dallas, and David Evans. 2003. "The Evolution of NOx Control Policy for Coal-Fired Power Plants in the United States." Discussion Paper 03-23. Washington: Resources for the Future (December).

Cain, Louis P., and Elyce J. Rotella. 2001. "Death and Spending: Urban Mortality and Municipal Expenditure on Sanitation." *Annales de Démographie Historique* 1: 139–54.

California Air Resources Board. 2004. *California Ambient Air Quality Data, 1980–2002*. Sacramento, Calif. CD-ROM.

Carson, Rachel. 1962. *Silent Spring*. Boston: Houghton Mifflin.

Chay, Kenneth Y., and Michael Greenstone. 2003. "The Impact of Air Pollution on Infant Mortality: Evidence from Geographic Variation in Pollution Shocks Induced by a Recession." *Quarterly Journal of Economics* 118, no. 3: 1121–67.

———. 2005. "Does Air Quality Matter? Evidence from the Housing Market." *Journal of Political Economy* 113, no. 2: 376–424.

Cohen, Joel. 1995. *How Many People Can the Earth Support?* New York: Norton.

Cohen, Linda, and Roger Noll. 1981. "The Economics of Disaster Defense: The Case of Building Codes to Resist Seismic Shock." *Public Policy* 29, no. 1: 1–29.

Collins, William, and Robert Margo. 2003. "Race and the Value of Owner-Occupied Housing." *Regional Science and Urban Economics* 33, no. 3: 255–86.

Copeland, Brian, and Scott Taylor. 2004. "Trade, Growth, and the Environment." *Journal of Economic Literature* 42, no. 1: 7–71.

Costa, Dora L., and Matthew E. Kahn. 2000. "Power Couples: The Locational Choice of the College Educated, 1940–1990." *Quarterly Journal of Economics* 115, no. 4: 1287–1315.

———. 2003a. "Civic Engagement and Community Heterogeneity: An Economist's Perspective." *Perspectives on Politics* 1, no. 1: 103–12.

———. 2003b. "The Rising Price of Nonmarket Goods." *American Economic Review* 93, no. 2: 227–32.

———. 2004. "Changes in the Value of Life, 1940–1980." *Journal of Risk and Uncertainty* 29, no. 2: 159–80.

Craddock, Susan. 2000. *City of Plagues: Disease, Poverty, and Deviance in San Francisco*. University of Minnesota Press.

Crandall, Robert W. 1993. *Manufacturing on the Move*. Brookings.

Cullen, Julie Berry, and Steven Levitt. 1999. "Crime, Urban Flight, and the Consequences for Cities." *Review of Economics and Statistics* 81, no. 2: 159–69.

Cutler, David M., Edward L. Glaeser, and Jacob L. Vigdor. 1999. "The Rise and Decline of the American Ghetto." *Journal of Political Economy* 107, no. 3: 455–506.

Cutler, David, and Grant Miller. 2005. "Water, Water, Everywhere: Municipal Finance and Water Supply in American Cities." Working Paper 11096. Cambridge, Mass.: NBER (January).

Daniere, Amrita G., and Lois M. Takahashi. 1997. "Environmental Policy in Thailand: Values, Attitudes, and Behavior among the Slum Dwellers of

Bangkok." *Environment and Planning C: Government and Policy* 15, no. 3: 305–27.

Dasgupta, Susmita, and others. 2002. "Confronting the Environmental Kuznets Curve." *Journal of Economic Perspectives* 16, no. 1: 147–68.

———. 2004. "Air Pollution during Growth: Accounting for Governance and Vulnerability." Policy Research Working Paper 3383. Washington: World Bank (August).

Deffeyes, Kenneth. 2001. *Hubbert's Peak: The Impending World Oil Shortage.* Princeton University Press.

Deichmann, Uwe, and Vernon Henderson. 2000. "Urban and Regional Dynamics in Poland." Policy Research Working Paper 2457. Washington: World Bank (September).

Diamond, Jared. 2005. *Collapse: How Societies Choose to Fail or Succeed.* New York: Viking.

DiPasquale, Denise, and Edward Glaeser. 1999. "Incentives and Social Capital: Are Homeowners Better Citizens?" *Journal of Urban Economics* 45, no. 2: 354–84.

DiPasquale, Denise, and Matthew E. Kahn. 1999. "Measuring Neighborhood Investments: An Examination of Community Choice." *Real Estate Economics* 27, no. 3: 369–424.

Ederington, Josh, Arik Levinson, and Jenny Minier. 2005. "Footloose and Pollution-Free." *Review of Economics and Statistics* 87, no. 1: 92–99.

Ederington, Josh, and Jenny Minier. 2003. "Is Environmental Policy a Secondary Trade Barrier? An Empirical Analysis." *Canadian Journal of Economics* 36, no. 1: 137–54.

Ehrlich, Paul. 1968. *The Population Bomb.* New York: Ballantine Books.

Ewing, Reid, and others. 2005. *Endangered by Sprawl: How Runaway Development Threatens America's Wildlife.* Washington: National Wildlife Federation, Smart Growth America, and NatureServe.

Feenstra, Robert C. 1996. "U.S Imports, 1972–1994: Data and Concordances." Working Paper 5515. Cambridge, Mass.: NBER. March.

Feinberg, Richard. 2005. "Book Reviews: Western Hemisphere." *Foreign Affairs* 85, no. 3.

Fischel, William A. 1999. "Does the American Way of Zoning Cause the Suburbs of Metropolitan Areas to Be Too Spread Out?" In *Governance and Opportunity in Metropolitan America*, edited by Alan Altschuler and others, pp. 151–91. Washington: National Academies Press.

Florida, Richard. 2002. *The Rise of the Creative Class: And How It's Transforming Work, Leisure, Community and Everyday Life.* New York: Basic Books.

Fogel, Robert W. 2000. *The Fourth Great Awakening and the Future of Egalitarianism.* University of Chicago Press.

Foster, Andrew, and Mark R. Rosenzweig 2003. "Economic Growth and the Rise of Forests." *Quarterly Journal of Economics* 118, no. 2: 601–37.

Fredriksson, Per, and Daniel Millimet. 2002. "Is There a 'California Effect' in U.S. Environmental Policymaking?" *Regional Science and Urban Economics* 32, no. 6: 737–64.

Fredriksson, Per, and others. 2005. "Environmentalism, Democracy and Pollution Control." *Journal of Environmental Economics and Management* 49, no. 2: 343–65.

Friedman, Michael S., and others. 2001. "Impact of Changes in Transportation and Commuting Behaviors during the 1996 Summer Olympic Games in Atlanta on Air Quality and Childhood Asthma." *JAMA* 285, no. 7: 897–905.

Fullerton, Don, and Thomas C. Kinnaman. 1996. "Household Responses to Pricing Garbage by the Bag." *American Economic Review* 86 (September): 971–84.

Gallagher, Kevin P. 2004. *Free Trade and the Environment: Mexico, NAFTA, and Beyond*. Stanford University Press.

Glaeser, Edward L., Joseph Gyourko, and Raven Saks. 2005. "Why is Manhattan So Expensive? Regulation and the Rise in House Prices." *Journal of Law and Economics* 48, no. 2: 331–70.

Glaeser, Edward, and Matthew E. Kahn. 2004. "Sprawl and Urban Growth." In *Handbook of Urban and Regional Economics*, vol. 4: *Cities and Geography*, edited by Vernon Henderson and Jacques-François Thisse, pp. 2481–2527. Amsterdam: Elsevier North-Holland.

Glaeser, Edward, Matthew E. Kahn, and Jordan Rappaport. 2000. "Why Do the Poor Live in Cities?" Discussion Paper 1891. Cambridge, Mass.: Harvard Institute of Economic Research (April).

Glaeser, Edward, Jed Kolko, and Albert Saiz. 2001. "Consumer City." *Journal of Economic Geography* 1, no. 1: 27–50.

Glaeser, Edward L., and others. 1992. "Growth in Cities." *Journal of Political Economy* 100, no. 6: 1126–52.

Goodstein, Eban. 1999. *The Trade-Off Myth: Fact and Fiction about Jobs and the Environment*. Washington: Island Press.

Gordon, Peter, Ajay Kumar, and Harry Richardson. 1991. "The Influence of Metropolitan Spatial Structure on Commuting Time." *Journal of Urban Economics* 26, no. 2: 138–51.

Greenstone, Michael. 2002. "The Impacts of Environmental Regulation on Industrial Activity." *Journal of Political Economy* 110, no. 6: 1175–1219.

Greenstone, Michael, and Justin Gallagher. 2005. "Does Hazardous Waste Matter? Evidence from the Housing Market and the Superfund Program." Working Paper 11790. Cambridge, Mass.: NBER.

Grossman, Gene, and Alan Krueger. 1995. "Economic Growth and the Environment." *Quarterly Journal of Economics* 110, no. 2: 353–77.

Gruenspecht, Howard. 1982. "Differentiated Regulation: The Case of Auto Emissions Standards." *American Economic Review* 72, no. 2: 328–31.

Haines, Michael. 2001. "The Urban Mortality Transition in the United States, 1800–1940." Historical Paper 134. Cambridge, Mass.: NBER (July).

Hamilton, James D. 1983. "Oil and the Macroeconomy since World War II." *Journal of Political Economy* 91, no. 2: 228–48.

Hammitt, James, Jin-Tan Liu, and Jin-Long Liu. 2000. "Survival Is a Luxury Good: The Increasing Value of a Statistical Life." Paper prepared for the Summer Institute Workshop on Public Policy and the Environment. Cambridge, Mass.: NBER (August).

Hanak, Ellen, and Ada Chen. Forthcoming. "Wet Growth: Effects of Water Policies on Land Use in the American West." *Journal of Regional Science.*

Hanemann, Michael. 2005. "The Economic Conception of Water." In *Water Crisis: Myth or Reality?* edited by Peter P. Rogers, M. Ramón Llamas, and Luis Martinez Cortina. Leiden, Netherlands: Taylor and Francis.

Harbaugh, William, Arik Levinson, and David Wilson. 2002. "Re-examining the Empirical Evidence for an Environmental Kuznets Curve." *Review of Economics and Statistics* 84, no 3: 541–51.

Hardin, Garrett. 1968. "The Tragedy of the Commons." *Science* 162, no. 3859: 1243–48.

Hazilla, Michael, and Raymond J. Kopp. 1990. "Social Cost of Environmental Quality Regulations: A General Equilibrium Analysis." *Journal of Political Economy* 98, no. 4: 853–73.

Henderson, Vernon. 1991. *Urban Development: Theory, Fact, and Illusion.* Oxford University Press.

———. 1996. "The Effect of Air Quality Regulation." *American Economic Review* 86, no. 4: 789–813.

———. 2002. "Urban Primacy, External Costs, and Quality of Life." *Resource and Energy Economics* 24, no. 1–2: 95–106.

Henderson, Vernon, Todd Lee, and Yung Joon Lee. 2001. "Scale Externalities in Korea." *Journal of Urban Economics* 49, no. 3: 479–504.

Henderson, J. Vernon, and Hyoung Gun Wang. 2004. "Urbanization and City Growth." Mimeo. Brown University.

Hilton, F. G. Hank, and Arik Levinson. 1998. "Factoring the Environmental Kuznets Curve: Evidence from Automotive Lead Emissions." *Journal of Environmental Economics and Management* 35, no. 2: 126–41.

Holmes, Thomas. 1998. "The Effect of State Policies on the Location of Manufacturing: Evidence from State Borders." *Journal of Political Economy* 106, no. 4: 667–705.

Hoy, Michael, and Emmanuel Jimenez. 1996. "The Impact on the Urban Environment of Incomplete Property Rights." Working Paper 14. Washington: World Bank (April).

Huang, Yan. 2004. "Urban Spatial Patterns and Infrastructure in Beijing." *Land Lines* 16, no. 4.

Hughes, Gordon, and Magda Lovei. 1999. "Economic Reform and Environmental Performance in Transition Economies." Technical Paper 446. Washington: World Bank.

International Federation of Red Cross and Red Crescent Societies. 2002. *World Disasters Report 2002: Focus on Reducing Risk.* Geneva.

Jackson, Kenneth. 1985. *Crabgrass Frontier: The Suburbanization of the United States.* Oxford University Press.

Jacobs, Jane. 1969. *The Economy of Cities.* New York, Vintage Press.

Jaffe, Adam B., Richard G. Newell, and Robert N. Stavins. 2002. "Environmental Policy and Technological Change." *Environmental and Resource Economics* 22, no. 1–2: 41–69.

Kahn, Matthew E. 1997. "Particulate Pollution Trends in the United States." *Regional Science and Urban Economics* 27, no. 1: 87–107.

———. 1999. "The Silver Lining of Rust Belt Manufacturing Decline." *Journal of Urban Economics* 46, no. 3: 360–76.

———. 2000. "The Environmental Impact of Suburbanization." *Journal of Policy Analysis and Management* 19, no. 4: 569–86.

———. 2001a. "Does Sprawl Reduce the Black/White Housing Consumption Gap?" *Housing Policy Debate* 12, no. 1: 77–86.

———. 2001b. "The Beneficiaries of Clean Air Act Legislation." *Regulation* 24, no. 1: 34–39.

———. 2002. "Demographic Change and the Demand for Environmental Regulation." *Journal of Policy Analysis and Management* 21, no. 1: 45–62.

———. 2003a. "New Evidence on Eastern Europe's Pollution Progress." *Topics in Economic Policy* 3, no. 1: article 4 (www.bepress.com/bejeap/topics/vol3/iss1/art4 [March 2006]).

———. 2003b. "The Geography of U.S. Pollution Intensive Trade: Evidence from 1958 to 1994." *Regional Science and Urban Economics* 33, no. 4: 383–400.

———. 2004. "Domestic Pollution Havens: Evidence from Cancer Deaths in Border Counties." *Journal of Urban Economics* 56, no. 1: 51–69.

———. 2005. "The Death Toll from Natural Disasters: The Role of Income, Geography and Institutions." *Review of Economics and Statistics* 87, no. 2: 271–84.

———. 2006. "Environmental Disasters as Regulation Catalysts: The Role of Bhopal, Chernobyl, *Exxon-Valdez*, Love Canal, and 3 Mile Island in Shaping Environmental Law." Working Paper. Tufts University.

Kahn, Matthew, and John Matsusaka. 1997. "Demand for Environmental Goods: Evidence from Voting Patterns on California Initiatives." *Journal of Law and Economics* 40, no. 1: 137–73.

Kahn, Matthew E., and Joel Schwartz. 2006. "Air Pollution Progress despite Sprawl: The 'Greening' of the Vehicle Fleet." Working paper. Tufts University.

Kain, John F. 1990. "Deception in Dallas: Strategic Misrepresentation in Rail Transit Promotion and Evaluation." *Journal of the American Planning Association* 56, no. 2: 184–96.

———. 1997. "Cost-Effective Alternatives to Atlanta's Rail Rapid Transit System." *Journal of Transport Economics and Policy* 31, no. 1: 25–49.

———. 1999. "The Urban Transportation Problem: A Re-examination and Update." In *Essays in Transportation Economics and Policy: A Handbook in Honor of John R. Meyer*, edited by Jose Gomez-Ibanez, William B. Tye, and Clifford Winston, pp. 359–401. Brookings.

Katakura, Yoko, and Alexander Bakalian. 1998. "PROSANEAR: People, Poverty and Pipes. A Program of Community Participation and Low-Cost Technology Bringing Water and Sanitation to Brazil's Urban Poor." Working paper. Washington: Water and Sanitation Program, World Bank (September).

Katz, Lawrence F., and Kenneth Rosen. 1987. "The Interjurisdictional Effects of Growth Controls on Housing Prices." *Journal of Law and Economics* 30, no. 1: 149–60.

Keller, Wolfgang, and Arik Levinson. 2002. "Pollution Abatement Costs and Foreign Direct Investment Inflows to U.S. States." *Review of Economics and Statistics* 84, no. 4: 691–703.

Kiel, Katherine, and Katherine McClain. 1995. "House Prices during Siting Decision Stages: The Case of an Incinerator from Rumor through Operation." *Journal of Environmental Economics and Management* 28, no. 2: 241–55.

Kirshen, Paul, and others. 2004. "Lack of Influence of Climate on Present Cost of Water Supply in the USA." *Water Policy* 6, no. 4: 269–79.

Klinenberg, Eric. 2002. *Heat Wave: A Social Autopsy of Disaster in Chicago.* University of Chicago Press.

Kohlhase, Janet. 1991. "The Impact of Toxic Waste Sites on Housing Values." *Journal of Urban Economics* 30, no. 1: 1–26.

Kotchen, Matthew, and Shawn Powers. Forthcoming. "Explaining the Appearance and Success of Voter Referenda for Open Space Conservation." *Journal of Environmental Economics and Management.*

Krugman, Paul, and Raul Livas. 1996. "Trade Policy and the Third World Metropolis." *Journal of Development Economics* 49, no. 1: 137–50.

Kuran, Timur, and Cass R. Sunstein. 1999. "Availability Cascades and Risk Regulation." *Stanford Law Review* 51 (April): 683–768.

La Porta, Rafael, and others. 1999. "The Quality of Government." *Journal of Law, Economics, and Organization* 15, no. 1: 222–79.

Lee, Chang-Moo, and Peter Linneman. 1998. "Dynamics of the Greenbelt Amenity Effect on the Land Market—The Case of Seoul's Greenbelt." *Real Estate Economics* 26, no. 1: 107–29.

Lee, Chulhee. 1997. "Socioeconomic Background, Disease, and Mortality among Union Army Recruits: Implications for Economic and Demographic History." *Explorations in Economic History* 34, no. 1: 27–55.

Levinson, Arik. 1996. "Environmental Regulations and Manufacturers' Location Choices: Evidence from the Census of Manufactures." *Journal of Public Economics* 62, no. 1-2: 5–29.

———. 2001. "An Industry-Adjusted Index of State Environmental Compliance Costs." In *Behavioral and Distributional Effects of Environmental Policy*, edited by Carlo Carraro and Gilbert Metcalf, pp. 131–58. University of Chicago Press.

Levitt, Steven D. 2004. "Understanding Why Crime Fell in the 1990s: Four Factors that Explain the Decline and Six That Do Not." *Journal of Economic Perspectives* 18, no. 1: 163–90.

Libecap, Gary D. 2005. "The Myth of Owens Valley." *Regulation* 28, no. 2: 10–17.

Lindert, Peter H. 1996. "What Limits Social Spending?" *Explorations in Economic History* 33, no. 1: 1–34.

Lomborg, Bjørn. 2001. *The Skeptical Environmentalist: Measuring the Real State of the World.* Cambridge University Press.

Luttmer, Erzo F. P. 2001. "Group Loyalty and the Taste for Redistribution." *Journal of Political Economy* 109, no. 3: 500–28.

Maddison, David, and Andrea Bigano. 2003. "The Amenity Value of the Italian Climate." *Journal of Environmental Economics and Management* 45, no. 2: 319–32.

Margo, Robert. 1992. "Explaining the Postwar Suburbanization of the Population in the United States: The Role of Income." *Journal of Urban Economics* 31, no. 3: 301–10.

McCully, Patrick. 1996. *Silenced Rivers: The Ecology and Politics of Large Dams*. London: Zed Books.

McMillen, Daniel, P. 2004. "Airport Expansions and Property Values: The Case of Chicago O'Hare Airport." *Journal of Urban Economics* 55, no. 3: 627–40.

Melosi, Martin. 1982. *Garbage in the Cities: Refuse, Reform, and the Environment: 1880–1980*. Texas A&M Press.

———. 2001. *Effluent America; Cities, Industry, Energy and the Environment*. University of Pittsburgh Press.

Mieszkowski, Peter, and Edwin S. Mills. 1993. "The Causes of Metropolitan Suburbanization." *Journal of Economic Perspectives* 7, no. 3: 135–47.

Miguel, Theodore, and Mary Kay Gugerty. 2005. "Ethnic Diversity, Social Sanctions, and Public Goods in Kenya." *Journal of Public Economics* 89, no. 11-12: 2325–68.

Moomaw, William R., and Gregory C. Unruh. 1997. "Are Environmental Kuznets Curves Misleading Us? The Case of CO_2 Emissions." *Environment and Development Economics* 2, no. 4: 451–63.

Moretti, Enrico. 2004. "Human Capital Externalities in Cities." In *Handbook of Urban and Regional Economics*, vol. 4, edited by Vernon Henderson and Jacques-François Thisse. Amsterdam: Elsevier North-Holland.

Neal, Derek. 1995. "Industry-Specific Human Capital: Evidence from Displaced Workers." *Journal of Labor Economics* 13, no. 4: 653–77.

Necheyba, Thomas J., and Randall Walsh. 2005. "Urban Sprawl." *Journal of Economic Perspectives* 18, no. 4: 177–200.

Neidell, Matthew. 2004. "Air Pollution, Health and Socio-Economic Status: The Effect of Outdoor Air Quality on Childhood Asthma." *Journal of Health Economics* 23, no. 6: 611–30.

Newell, Richard, Adam Jaffe, and Robert Stavins. 1999. "The Induced Innovation Hypothesis and Energy-Saving Technological Change." *Quarterly Journal of Economics* 114, no. 3: 941–75.

Newman, Peter, and Jeffrey Kenworthy. 1999. *Sustainability and Cities: Overcoming Automobile Dependence*. Washington: Island Press.

Nivola, Pietro. 1999. *Laws of the Landscape: How Policies Shape Cities in Europe and America*. Brookings.

Nivola, Pietro, and Robert Crandall. 1995. *The Extra Mile*. Brookings.

Noll, Roger, Mary M. Shirley, and Simon Cowan. 2000. "Reforming Urban Water Systems in Developing Countries." In *Economic Policy Reform: The Second Stage*, edited by Anne O. Krueger. University of Chicago Press.

Nordhaus, William D. 1992. "Lethal Model 2: The Limits to Growth Revisited." *BPEA*, no. 2: 1–59.

Nordhaus, William D., and Joseph Boyer. 2003. *Warming the World: Economic Models of Global Warming.* MIT Press.

Ofek, Haim, and Yesook Merrill. 1997. "Labor Immobility and the Formation of Gender Wage Gaps in Local Markets." *Economic Inquiry* 35, no. 1: 28–47.

Olson, Mancur. 1965. *Logic of Collective Action: Public Goods and the Theory of Groups.* Harvard University Press.

O'Toole, Randall. 2001. "The Folly of Smart Growth." *Regulation* 24, no. 3: 20–26.

Otto, Betsy, and others. 2003. *Paving Our Way to Water Shortages: How Sprawl Aggravates Drought.* Washington: American Rivers, National Resource Defense Council, and Smart Growth America.

Palmer, Margaret, and others. 2002. "The Ecological Consequences of Changing Land Use for Running Waters, with a Case Study of Urbanizing Watersheds in Maryland." *Yale School of Forestry and Environmental Studies Bulletin*, no. 107: 85–113.

Panayotou, Theo. 2000. "Globalization and the Environment." Working Paper 53. Cambridge, Mass.: Center for International Development, Harvard University.

Parikh, Jyoti, and Vibhooti Shukla. 1995. "Urbanisation, Energy Use and Greenhouse Effects in Economic Development: Results from a Cross-National Study of Developing Countries." *Global Environmental Change* 5, no. 2: 87–105.

Parry, Ian W. H., and Kenneth A. Small. 2005. "Does Britain or the United States Have the Right Gasoline Tax?" *American Economic Review* 95, no. 4: 1276–89.

Pfaff, Alexander S. P., Shubham Chaudhuri, and H. Nye. 2004. "Household Production and Environmental Kuznets Curves: Examining the Desirability and Feasibility of Substitution." *Environmental and Resource Economics* 27, no. 2: 187–200.

Phillips, Justin, and Eban Goodstein. 2000. "Portland's Urban Growth Boundary." *Contemporary Economic Policy* 18, no. 3: 334–44.

Popp, David. 2002. "Induced Innovation and Energy Prices." *American Economic Review* 92, no. 1: 160–80.

Portney, Kent. 2003. *Taking Sustainable Cities Seriously: Economic Development, the Environment, and Quality of Life in American Cities.* MIT Press.

Ransom, Michael, and C. Arden Pope. 1995. "External Health Costs of a Steel Mill." *Contemporary Economic Policy* 13, no. 2: 86–97.

Rappaport, Jordan. 2003. "Moving to Nice Weather." Research Working Paper RWP 03-07. Federal Reserve Bank of Kansas City.

Rauch, James E. 1993. "Productivity Gains from Geographic Concentration of Human Capital: Evidence from the Cities." *Journal of Urban Economics* 34, no. 3: 380–400.

Rawls, John. 1999. *A Theory of Justice.* Revised ed. Harvard University Press.

Schmalensee, Richard, Thomas Stoker, and Ruth Judson. 1998. "World Carbon Dioxide Emissions: 1950–2050." *Review of Economics and Statistics* 80, no. 1: 15–28.

Selden, Thomas M., and Daqing Song. 1995. "Neoclassical Growth, the J Curve for Abatement, and the Inverted U Curve for Pollution." *Journal of Environmental Economics and Management* 29, no. 2: 162–68.

Seroa da Motta, Ronaldo, and Leonardo Rezende. 1999. "Estimation of Water Quality Control Benefits and Instruments in Brazil. The Impact of Sanitation on Waterborne Diseases in Brazil." In *Natural Resource Valuation and Policy in Brazil*, edited by Peter H. May. Columbia University Press.

Sigman, Hilary. 2001. "The Pace of Progress at Superfund Sites: Policy Goals and Interest Group Influence." *Journal of Law and Economics* 44, no. 1 (April): 315–44.

———. 2002. "International Spillovers and Water Quality in Rivers: Do Countries Free Ride?" *American Economic Review* 92, no. 4: 1152–59.

Soto, Hernando de. 2000. *The Mystery of Capital: Why Capitalism Triumphs in the West and Fails Everywhere Else*. New York: Basic Books.

Stavins, Robert. 1992. "Comments on 'Lethal Model 2: The Limits to Growth Revisited' by William D. Nordhaus." *BPEA*, no. 2: 44–50.

———. 1998. "What Can We Learn from the Grand Policy Experiment? Lessons from SO_2 Allowance Trading." *Journal of Economic Perspectives* 12, no. 3: 69–88.

Timmins, Christopher. 2002. "Does the Median Voter Consume Too Much Water? Analyzing the Redistributive Role of Residential Water Bills." *National Tax Journal* 55, no. 4: 687–702.

Troesken, Werner. 2004. *Water, Race, and Disease*. MIT Press.

U.S. Environmental Protection Agency. 2000. *A Benefits Assessment of Water Pollution Control Programs since 1972. Part 1: The Benefits of Point Source Controls for Conventional Pollutants in Rivers and Streams*. Washington: Office of Policy, Economics and Innovation (www.epa.gov/ost/economics/assessment.pdf [May 2006]).

———. 2004. *The Particle Pollution Report: Current Understanding of Air Quality and Emissions through 2003*. Research Triangle Park, N.C.: Office of Air Quality Planning and Standards.

Veblen, Thorstein. 1899. *The Theory of the Leisure Class: An Economic Study of Institutions*. New York: Macmillan.

Vernon, Raymond. 1964. "The Myth and Reality of Our Urban Problem." In *City and Suburb: The Economics of Metropolitan Growth*, edited by Benjamin Chinitz. Englewood Cliffs, N.J.: Prentice Hall.

Vigdor, Jacob. 2004. "Community Composition and Collective Action: Analyzing Initial Mail Response to the 2000 Census." *Review of Economics and Statistics* 86, no. 1: 303–12.

Viscusi, W. Kip. 1993. "The Value of Risks to Life and Health." *Journal of Economic Literature* 31, no. 4: 1912–46.

Vries, Jaap de, and others. 2001. "Environmental Management of Small and Medium Sized Cities in Latin America and the Caribbean." Working paper. Washington: Institute for Housing and Urban Development Studies, Inter-American Development Bank (January).

Wackernagel, Mathis, and others. 2002. "Tracking the Ecological Overshoot of the Human Economy." *Proceedings of the National Academy of Science* 99, no. 14: 9266–71.

Wilson, William J. 1990. *The Truly Disadvantaged: The Inner City, the Underclass, and Public Policy.* University of Chicago Press.

Wong, Grace. 2005. "Has Sars Infected the Housing Market? Evidence from Hong Kong." Working paper. Wharton School, University of Pennsylvania (March).

World Bank. 2001. *World Development Indicators 2001.* Washington.

Index

Accountability, 69
Accra (Ghana), 30
Acid rain, 47
Ades, Alberto, 99
Afghanistan, 97
Africa, 32, 97
Air conditioning, 38, 39, 112
Air quality/pollution. *See* Environmental issues
Air travel, 35
Alaska, 72
American Dream, 6, 10
American Farmland Trust, 60
American Housing Survey (*1999*), 121–22
Angola, 68
Argentina, 99
Ashenfelter, Orley, 16
Asia, 1, 97
Atlanta (GA), 14–15, 113, 114, 120
Automobiles and automobile industry: air pollution and, 76; commuting in, 118, 121; economic factors, 31; effects of sprawl on, 113–21; gasoline and oil and, 11–12, 75; hybrid vehicles, 38, 39, 51, 57, 58; industrial waste, 63; regulation and prices, 92; smog and, 18n24; SUV ownership, 116; vehicle emissions, 52, 57, 58, 67, 75–76. *See also* Gasoline; Oil

Baltimore (MD), 120
Bangkok (Thailand), 30, 105
Baum-Snow, Nathaniel, 112
Beijing (China), 36
Bhagwati, Jagdish, 48
Bhopal (India), 42
Bienenfeld, Robert, 58
Big Dig (Boston, MA), 89–90
Black Plague, 99
Boston (MA): Big Dig, 89–90; employment in, 111; housing supply and regulation in, 24; public transit in, 118, 120, 121; vehicle use in, 113, 114
Boulder (CO), 88, 126
Boyer, Joseph, 137
Brazil, 97, 103, 104
Brownfields, 90, 91, 125
Bureau of Automotive Repair, 52
Bureau of Transportation Statistics, 53
Bush, George W., 90
Bush (George W.) administration, 135
Business and corporate issues: business improvement districts, 70; corporate location, 111–12; environmental regulation, 73; pollution and pollu-

tion abatement, 59–60, 70–71, 74n21, 78; regulatory grandfathering, 78–79; Toxic Release Inventory, 92. *See also* Superfund Program

Cain, Louis, 83–84
California: air pollution in, 75–76, 80–83; automobile emissions, 52, 53t, 75–76, 77f; Clean Air Act and, 80–81; education and green policies, 71; gasoline tax proposition (*185*), 119–20; MTBE in, 79; quality of life in, 21n29; regulation in, 24, 46; water issues in, 107, 124–25. *See also* Marin County; *individual cities*
California Air Resources Board, 80
California Coastal Commission, 73n14
California Trust for the Public Land, 60
Canada, 33, 68, 135
Cap and trade systems, 135
Carbon dioxide: from burning gasoline, 114n18; cap and trade system for, 135; ecological footprint and, 4, 26, 27; economic factors, 38, 48; EKC and, 134; fossil fuels and, 10n5; production of, 135; Regional Greenhouse Gas Initiative, 135n11; sprawl and, 110; value of, 28–29
Carbon monoxide, 52, 76, 80, 81
Cars. *See* Automobile industry
Census of Population and Housing (*2000*; U.S.), 21, 118
CERCLA. *See* Comprehensive Environmental Response, Compensation, and Liability Act
Charlotte (NC), 114
Charlottesville (VA), 87
Chay, Kenneth, 15, 23
Chicago (IL): air pollution in, 1, 15; climate premium of, 23; heat wave of *1995*, 133; metropolitan area of, 3; noise pollution in, 35; noxious facilities in, 91n62; population of, 128; public transit in, 118, 120
China, 2, 32–33, 36, 47
Chinatown (San Francisco, CA), 129n54

Cholera, 37
Cities: brown cities, 13–14, 20, 27, 108, 126, 130, 131; center cities, 110, 111, 112, 121, 128, 129; correcting and cleaning past mistakes, 89–91; definition of, 3; diversity and growing cities, 106–09; ecological footprint of, 4, 10–12, 26, 29; economic issues, 3, 36–37, 74, 98, 101, 126; environmental degradation or improvement, 1–2, 3, 5, 100–106; expensive and cheap real estate in, 20–26; gentrification, 79–80; green cities, 3–5, 8, 17, 20, 26–29, 65, 67, 68, 85, 105, 114n18, 122n35, 125–26, 130, 131; manufacturing in, 62, 64n24; megacities, 97–99; noisy cities, 35; population of, 2, 10–11, 93–106, 126, 131; pollution permit systems, 78; quality of life in, 23, 25, 26, 31n1, 74, 126–27, 130, 136; smart growth cities, 124t, 125–28; sprawling cities, 114n18. *See also* Metropolitan areas; Urban areas; *individual cities*
City of London (Various Powers) Act of *1954*, 40
Civic engagement, 108–09
Clean Air Act (U.S.; *1970*), 23, 67, 75, 77–83, 92
Clean Air Acts (UK; *1956*, *1968*), 40
Clean Water Act (U.S.; *1972*), 83
Cleveland (OH), 63
Climate and climate change: benefits of, 26; city growth and, 113, 123; effects of, 7, 17–19, 25, 29, 101, 109; global temperature changes, 132–34; greenhouse gases and, 47, 132; as a political issue, 43; as an urban amenity, 21–23, 128; urban growth and, 134; water prices and, 123. *See also* Environmental issues
Clinton (Bill) administration, 33, 135
Collapse (Diamond), 10
Columbia University, 67–68
Communism, 64, 65
Compensating differentials. *See* Economic issues

INDEX   153

Comprehensive Environmental Response, Compensation, and Liability Act (CERCLA; *1980*), 90. *See also* Superfund Program
Conservation Fund, 60
Consumer issues: conspicuous consumption, 58; demand for green, 51–56; greening urban consumption, 50, 58–59; paying more for green, 56–58; supply of green, 59–61
Costa, Dora, 23
Counties, 15, 61, 79, 88, 102, 127. *See also* King County; Marin County
Crime, 23, 111
Cuyahoga River (Ohio), 39–40, 63
Czech Republic, 64

Dallas (TX), 111, 117, 120
Dasgupta, Susmita, 17–18
"Day of shame," 42
D.C. (District of Columbia). *See* Washington
Denver (CO), 18n24, 85, 120
Detroit (MI), 21, 63, 112
Developed countries, 97, 99. *See also individual countries*
Developing countries: demand for natural capital in, 10; indoor air pollution, 59; land management in, 89; move of dirty production to, 44; pollution in, 6; population and the environment in, 93; value of statistical life in, 17. *See also individual countries*
Diamond, Jared, 10
Diet, 51, 58
DinAlt, Jason, 106
DiPasquale, Denise, 23
Disasters, 37
District of Columbia (D.C.). *See* Washington

Earth Day International, 9
Earthquakes, 37
Ecological footprint (ecofootprint). *See* Environmental issues
Ecological footprint calculator, 9–10

Economic issues: air quality regulation, 78–79; capital markets, 103–04; compensating differentials, 21, 24–26; conservation and prices, 49; costs of pollution, 15–17; costs of reducing air pollution, 28, 29, 82–83; dollar value of statistical life, 16, 28; environmental quality, 3, 4, 43–44, 50–60; foreign direct investment, 44; garbage disposal, 86–87; implicit price of climate, 23; income redistribution, 108; local health levels, 19–20; market incentives, 135; market supply, demand, and prices, 11–12, 20–21, 24, 51–60; market trading systems, 135; natural capital, 8, 9, 10, 11, 12, 28, 48, 73n14, 104, 122, 123; "peak toll" pricing, 85n48; per capita gross domestic product (GDP), 32, 36; per capita gross national product (GNP), 34; per capita income/wealth, 31, 32, 33, 35–36, 37n16, 39, 44, 48, 50–58, 67–68, 71–74, 80, 103, 107, 113, 118, 119, 126, 132; poverty and the poor, 19, 29, 58–59, 104, 119, 126, 133; productivity, 92; real estate prices, 4, 8, 20–26, 91n60, 127; recessions, 15, 38; regulation, 78, 92; research and development, 38; revealed preferences, 24, 28; trade, 33–34, 44, 45–46, 62, 99–100; urban growth, 6, 19–20. *See also* Employment issues; Environmental Kuznets curve
Economist, The (newsmagazine), 65–66
Educational issues: economic factors, 67; educational level and demand for green policies, 51, 71–72, 126; educational level and manufacturing jobs, 62–63; investment in education, 108; migration to the suburbs, 111
Ehrlich, Paul, 13
EKC. *See* Environmental Kuznets curve
Electric utilities, 75, 78
Eminent domain, 60
Employment issues: dirty jobs, 45;

employment sprawl and suburbanization, 111–12, 121; manufacturing jobs, 61, 62–63; megacity labor force, 98; risk and risky jobs, 16; service sector jobs, 61–63; wages, 16, 28; women in the labor force, 98, 134

Energy issues: content of manufactured goods, 45–46; dirty fuels, 59, 64; fossil fuels, 10, 12n10; energy prices, 38; impending energy crisis, 13

Environmental issues: air quality and pollution, 15n16, 75–83, 100–102, 128; available data, 32; conservation, 85, 124; cross-boundary externalities, 47; ecological footprints, 4, 8–13, 27, 28, 29, 48, 71n8, 85, 106, 107, 114n18, 123, 127–28, 130, 131; emissions controls, 51; environmental justice, 80–83, 91n62; environmental morbidity, 27, 29, 32–33; environmental mortality, 27–28, 29, 32–33, 83–84, 103, 133; externalities, 3, 5, 43, 47, 69, 88, 123, 132; global consumption, 10; greenhouse gases, 12, 47, 48, 68, 114, 128, 134; lead pollution and exposure, 35–36, 43, 51; media coverage of, 41–42, 43; most important issues, 72; NAFTA, 33; pollution, 14–17, 27, 30–31, 32–33, 35, 43, 44–47; population growth, 100–106; resource supply and demand, 12–13; technology, 65–66; water quality, 63, 83–85, 102–04, 116. *See also* Climate and climate change; Public health issues; Smog; *individual pollutants*

Environmental Kuznets curve (EKC): data on emissions and miles driven, 53–56; examples of, 30, 35–37; hypothesis of, 3, 5, 6, 30–33, 48, 130–31; limitations of, 131–32; objections to, 3, 43–47; origins of, 33–34, 50; regulation and, 49, 67, 92; shifting of, 37–43, 50; turning point of, 30, 32, 34, 35, 36–37, 38, 39, 44, 65, 130–31, 132, 134; urban sprawl, 110–29; when the EKC appears, 48–49

Environmental Protection Agency (EPA), 42, 85, 91, 102

Environmental Sustainability Index (ESI), 67–68

EPA. *See* Environmental Protection Agency

Erie (PA), 63

ESI. *See* Environmental Sustainability Index

Europe, 64–65, 75, 97, 99, 108. *See also individual countries*

Exxon Valdez oil spill (1989), 39–40

Farms and farmland, 122, 123–24, 127n49, 133–34

FEC. *See* Federal Election Commission

Federal-Aid Highway Act of *1956*, 112

Federal Election Commission (FEC), 71

Federal Highway Administration (FHA), 53

Finland, 68

Fischel, William, 127

Food, 56–57

France, 97

Fredriksson, Per, 36

Free-rider problems, 57, 70, 88, 108

Fresh Kills garbage dump (NYC), 86

Fullerton, Don, 87

Gambling and gamblers, 16

Gary (IN), 63

Gasoline: conservation of, 38; consumption of, 113–16, 128; hybrid cars and, 57–58; lead content of, 35–36; price of, 12; taxes on, 75, 119–20

GEMS. *See* Global Environmental Monitoring System

Geographic factors, 17–19, 26, 109

Gilmore, Jim, 87

Giuliani, Rudy, 87

Glaeser, Edward, 24, 99

Global Environmental Monitoring System (GEMS), 32, 34

Globalization, 64n24

Global sustainability, 130–37

Global warming. *See* Climate and climate change
Gloversville (NY), 98
Gore, Albert, Jr., 33–34
Government and governance: air quality and, 101–02; demand for green policies, 70–74; of diverse cities, 106–07; necessity for government interventions, 68–70; role of, 12, 39–43, 88; supplying greener governance, 74–91. *See also* Regulation
Great Lakes, 90–91. *See also* Lake Erie; Lake Michigan
Great Smog (London; 1952), 40–41
Green cities. *See* Cities
Green city index, 26–29, 48, 122n35
Greenhouse gases. *See* Environmental issues; *individual gases*
Greenness, 4, 50–60
Greenstone, Michael, 15, 16, 23
Grossman, Gene, 33, 34, 48
Gugerty, Mary Kay, 108
Gulf of Mexico, 72

Haiti, 68
Hanemann, Michael, 49
Harbaugh, William, 34
Hazardous waste, 23, 42. *See also* Solid waste
Henderson, Vernon, 97
Highways and roads, 35–36, 108, 110, 116, 119. *See also* Transportation
Homeownership, 105. *See also* Housing
Hong Kong, 25, 47, 65–66
Housing, 24, 37n16, 58, 72–73, 79–80, 91n60, 127
Houston (TX), 21, 24, 114
Hubbert, M. King, 12
Hubbert's peak, 12–13
Hungary, 64
Hurricanes, 12, 25n40
Hydrocarbons, 52, 53–56, 75
Hynes, John, 89–90
Hypotheses and models: effects of deindustrialization, 64; factor endowment hypothesis, 46n37; free-rider hypothesis, 57–58; induced innovation hypothesis, 38; pollution havens hypothesis, 44–47; relationship between economy and regulation, 39; relationship between education and environmentalism, 126. *See also* Environmental Kuznets curve; Methods and models

Iceland, 68
Immigration and migration: area desirability and, 25; immigration to the U.S., 106; of manufacturing, 64; regulations and, 79; to richer nations, 106; sprawl and, 110; technology and, 112; urban in-migration, 11, 21, 131, 133–34; urban out-migration, 99
India, 17, 32–33, 37, 47, 59
Indonesia, 32
Information: alerts and announcements, 18; asymmetries of, 69, 92; concealing of, 59; declining resources and, 12n10; in developing countries, 89; economic effects of, 80n34; education and, 51; EKC and, 39–40, 49; importance of, 42; public health and, 20n28; technology and, 135n10
Infrastructure: capital markets and, 103; financing of, 107, 109; investment in, 85; migration and, 93, 133–34; public health infrastructure, 20n28; rail infrastructure, 117; urban growth and, 66. *See also* Highways and roads; Sewage and sewers
Inland Steel Company, 63
International Development Bank, 105
International Panel on Climate Change, 132
Internet, 18
Interstate highway system, 112
Iraq, 68
Italy, 21–22, 97

Jacobs, Jane, 106
Japan, 68
Judicial systems, 74

Kenya, 108. *See also* Africa
King County (WA), 88
Kinnaman, Thomas, 87
Korea. *See* North Korea; South Korea
Kotchen, Matthew, 88–89
Krueger, Alan, 33, 34, 48
Kuran, Timur, 42
Kuznets, Simon, 3n6. *See also* Environmental Kuznets curve

Labor and union issues, 47
Lake Erie, 63. *See also* Great Lakes
Lake Michigan, 63. *See also* Great Lakes
Land and land consumption, 121–23
Landfills/garbage dumps. *See* Solid waste
Land Trust Alliance, 60
Las Vegas (NV), 25, 112, 123
Latin America, 99, 105. *See also individual countries*
League of Conservation Voters (LCV), 47, 72
Levinson, Arik, 34
Liberia, 68
Life (statistical value of), 16–17
Lindert, Peter, 108
Lomborg, Bjorn, 43
London (UK), 18n24, 40–41, 62, 133. *See also* United Kingdom
Los Angeles (CA): air pollution in, 15, 101–02; employment in, 111; light-rail system in, 120; population of, 99, 102; real estate prices in, 21; smog in, 1–2, 18, 79; water issues in, 49, 84, 125
Love Canal (NY), 42

Manhattan. *See* New York City
Manufacturing and manufactured goods, 45–47, 48, 59–60, 61–65, 89
Marin County (CA), 127–28. *See also* California; San Francisco
Mass transit, 39, 110. *See also* Public transit; Transportation
Media, 41–42, 59, 71
Melosi, Martin, 128–29

Methods and models: building a green city index, 27–29; calculation of average population exposure, 97; calculation of gasoline usage, 114; calculation of percent commute using public transit, 118; calculation of pollution exposure, 80–81; calculation of population growth, 113n11; hedonic methods, 25n38, 28, 70n6, 86; measuring pollution's costs, 15–17; measuring pollution's effects, 14–15; measuring regulation's effects on pollution, 70; public health methods, 70n6; relationship between pollution and GNP, 34. *See also* Hypotheses and models
Methyl tertiary-butyl ether (MTBE), 79
Metropolitan areas: commuting and transportation in, 111; definition of, 3–4; gasoline consumption in, 113–16; growth of, 6; housing supply conditions in, 24; population in, 110. *See also* Cities; Urban areas; *individual areas and cities*
Mexico, 33
Mexico City (Mexico), 1, 4, 17, 18
Miami (FL), 25, 120
Microsoft, 112
Migration. *See* Immigration and migration
Miguel, Edward, 108
Money magazine, 28
Moscow (Russia), 22
MTBE. *See* Methyl tertiary-butyl ether

NAFTA. *See* North American Free Trade Agreement
National Ambient Air Quality Standards, 77, 83n41
National Household Travel Survey, 53–56, 114
National Personal Transportation Survey, 113
National Priority List, 23
National Resources Defense Council, 116
Nature Conservancy, 60

New Jersey, 123
New Source Performance Standards, 75
New York City (NY): air pollution in, 30, 63; commuting in, 117, 118; employment downtown, 111; environment of, 1; garbage disposal in, 86–87; gasoline consumption in, 114; housing supply in, 24; population of, 99, 128; smog in, 18n24; water quality in, 84; zoning in, 89
New York County (Manhattan; NY), 61, 62
New York Times, 13, 41–42, 43
Nitrogen dioxide, 76, 77, 80, 82
Nitrogen (nitric and nitrous) oxides, 18n24, 52, 75, 77, 81
Noise, 35, 43, 47, 129, 131
Nordhaus, William, 134, 137
North American Free Trade Agreement (NAFTA), 33
Northeast (U.S.), 24, 25–26
North Korea, 68
Norway, 68

OECD. *See* Organization for Economic Cooperation and Development
O'Hare Airport (Chicago, IL), 35
Oil, 11–13, 38, 79
Oil Drum (blog site), 13
Olympic Games (*1996*), 14–15
OPEC. *See* Organization of Petroleum Exporting Countries
Open space initiatives, 88–89, 127–28
Organic foods. *See* Foods
Organic Foods Production Act (U.S.; *1990*), 67
Organization for Economic Cooperation and Development (OECD), 18, 108
Organization of Petroleum Exporting Countries (OPEC), 38
Orlando (FL), 114
Ozone: in California, 76, 81–82; effects of, 17, 23; formation of, 18n24; in Los Angeles, 18, 101–02; during the Olympic Games, 14–15; regulation of, 15n16; in U.S. cities, 77

Paris (France), 30
Particulates and particulate matter: in Asia, 1; in California, 82t; data on, 32, 80; definition of, 15n18; deindustrialization and, 63; effects of, 15, 17; EKC and, 34, 65; in eastern Europe, 64; in Hong Kong, 47; during the Olympic Games, 15; populations and, 100–101, 102; real estate prices and, 23; in the U.S., 76–77
Petroleum. *See* Oil
Philadelphia (PA), 112
Phoenix (AZ), 112, 114, 123
Pittsburgh (PA), 1, 15, 21, 63, 114
Poland, 64, 65
Political issues: campaign contributions, 71; climate change, 135; elections, 70; federal funding for infrastructure projects, 85; green/pro-environmental policies, 5, 47, 48–49, 70–72, 74, 107–08; land consumption, 123; landfills/garbage dumps, 73; "peak toll" pricing, 85n48; public political participation, 71; quality of life, 129; resource shortages, 11n7; suburbanization, 111n6; transportation, 110, 119; urban sustainability, 67; water supply and demand, 124–25
Pope, C. Arden, 14
Population, 93–100. *See also* Cities; Particulates and particulate matter; Urban areas; *individual cities*
Portland (OR), 4, 120, 126, 127
Portney, Kent, 125–26
Power lines, 24–25
Powers, Shawn, 88–89
Property rights, 69, 73n14, 73n16, 105
Proposition *185* (CA), 119–20
Public goods, 108
Public health issues: alerts and announcements, 18–19; costs of environmental problems, 15–17; economic factors, 37, 58, 130; environmental factors, 4, 8, 71n8, 130; income and diet, 20n28; infant mortality, 15; lead exposure, 35–36, 51; megacity growth, 99; pollution and

city health, 17; population density, 128–29; protective strategies, 17–20; service provision, 102–03, 128–29; solid waste, 86, 105; urban environmental quality, 13–20, 26; water pollution, 83–84. *See also* Environmental issues; Love Canal
Public transit, 107, 108, 116–21, 126. *See also* Mass transit

Racial, minority, and ethnic issues, 80, 82, 91n62, 108, 127, 129
Rail transit, 117, 120–21
Ransom, Michael, 14
Rawls, John, 19
Recycling (solid waste), 87
Redefining Progress, 9
Regulation: of air quality, 75–83; costs of, 78, 91–92; demand for green policies, 70–74; development and, 31; economic issues, 5, 39, 44, 45, 46n37, 68, 131; EKC and, 35, 49, 67; enforcement of, 91; environmental justice and, 80–83; grandfathering of, 78–79; Great Smog of London and, 40; housing and, 24; information regulation, 18; media and, 41–42; NAFTA and, 33; need for, 68–70; political factors of, 39, 42–43, 49; pollution and, 46, 78; state regulations, 46, 47, 72; support for, 51, 132; of water quality, 83–85; zoning and building codes, 37n16. *See also* Government and governance
Research and development, 38, 78
Richmond (WA), 112
Roads. *See* Highways and roads
Rotella, 83–84
Rural areas: consumption in, 11; life expectancy in, 2, 83; manufacturing in, 62, 63, 79; rural to urban migration, 11, 106, 133–34; solid waste disposal in, 87
Russia, 21–22
Rust Belt (U.S.), 15, 62–63

Sacramento (CA), 114, 120
San Antonio (TX), 114

San Diego (CA), 113, 120
San Francisco (CA): commuting in, 127–28; gasoline consumption in, 114; home prices in, 4; public health issues, 20n28; public transit use in, 117, 118, 120; real estate prices in, 21, 23, 24n37; sustainability of, 126. *See also* California; Marin County
Sanitation, 102–03. *See also* Sewage and sewers
San Jose (CA), 126
Santa Monica (CA), 79, 84–85, 126
Santa Monica Urban Runoff Recycling Facility (SMURRF), 84
Santiago (Chile), 17
SARS. *See* Severe acute respiratory syndrome
Saudi Arabia, 11n8
Scarsdale (NY), 111
Scottsdale (AZ), 126
Seattle (WA), 88, 126
Services, 61–63, 102–03, 107, 111
Severe acute respiratory syndrome (SARS), 25
Sewage and sewers, 23, 83–84, 98, 104, 108, 128, 129. *See also* Infrastructure
Siberia, 22
Sigman, Hilary, 91, 104
Simon, Julian, 13
Skeptical Environmentalist, This (Lomborg), 43
Small Business Liability Relief and Brownfields Revitalization Act (2002), 90
Smart growth, 60, 124t, 125–28
Smog, 18–19, 23, 79, 80, 101–02. *See also* Environmental issues
Smoking, 51
SMURRF. *See* Santa Monica Urban Runoff Recycling Facility
Social programs, 108
Solid waste, 73, 85–88, 104–05, 108. *See also* Hazardous waste
Soto, Hernando de, 105
South (U.S.), 63
South Korea, 47, 63
Species extinction, 123

Sprawl. *See* Environmental issues; Suburban areas; Urban areas
Statewide Database (University of California), 119
Statewide Emissions Inventory Summary (CA; *1995*), 76
Steel industry, 63
Suburban areas: businesses in, 111–12; commuting and driving in, 111, 112, 113, 116–18, 121; effects of suburbanization, 129, 131–32; gasoline consumption in, 128; land consumption in, 121–23; migration to, 111; smart growth in, 60; sprawl in, 6, 110, 113–25, 128, 131; urban land management and, 88
Sulfur dioxide: in California, 76; in eastern Europe, 64; EKC and, 34; global levels of, 101; as a measure of air pollution, 15n16; population and, 102n10; trading market for, 78, 135n10; in U.S. cities, 77
Sun Belt (U.S.), 112–13
Sunstein, Cass, 42
Superfund National Priority List, 91
Superfund Program, 23, 42, 90, 91. *See also* Comprehensive Environmental Response, Compensation, and Liability Act
Sweden, 68
Switzerland, 68

Taiwan, 68
Tax issues: carbon taxes, 12; gasoline taxes, 75, 119–20; incentives, 126; sales taxes, 88; service funding, 86–87, 111; tax base of center cities, 111n6
Technological issues: climate change and, 134; conservation, 107; EKC, 38–39, 131; environmental quality, 65–66, 102, 131, 133; green technologies, 11–12, 38–39, 48, 63, 65–66, 124–25; migration and, 112; regulation, 75–76; resource supply and demand, 13
Texas, 46–47
Theory of the Leisure Class, The (Veblen), 58

Third World. *See* Developing countries
Time magazine, 42
Timmins, Christopher, 107, 125
Toledo (OH), 63
Tourism, 74, 90, 121, 133
Toxic Release Inventory (TRI; U.S.), 42, 92
Trade. *See* Economic issues
Tragedy of the commons, 12n10, 69, 132, 135n10
Transportation, 62, 112. *See also* Highways and roads
Transportation, Department of (U.S.), 53, 114
TRI. *See* Toxic Release Inventory
Triangle Shirtwaist Fire (NYC; *1911*), 89
Troesken, Werner, 129
Turkmenistan, 68
Two Forks dam (CO), 85

UK. *See* United Kingdom
Union Carbide plant disaster (Bhopal, India), 42
United Kingdom (UK), 97. *See also* London
United Nations Environment Programme, 132
United States (U.S.): air quality in, 76–77; carbon dioxide emissions in, 26, 135; disasters in, 37; environment in, 106, 137; ESI ranking, 68; gasoline taxes, 75; global warming in, 137; greenhouse gases in, 114; life expectancy in, 2; manufacturing in, 61–62; miles driven in, 113; monitoring of air quality, 32; NAFTA and, 33; peak oil production in, 12; population in, 113; segregation in, 80; torts system in, 74; value of statistical life in, 16; views of global climate change in, 135–36; water in, 83–84, 123; welfare in, 108. *See also individual states*
Urban areas: climate change in, 7; commuting in, 116–18; consumption in, 11, 106; diversity and growth in, 106–09; economic development of, 2, 5, 7, 131, 132; green-

ing production in, 61–65; health planning, 17; land management in, 88–91; pollution strategies in, 19–20; population in, 6, 93–106, 110, 128, 131, 132; quality of life in, 24–26, 29, 67, 132, 137; spatial growth and sprawl of, 6, 29, 131; technology and environmental quality, 65–66, 136; urban consumption, 50–60; urban runoff, 84–85, 116; urban sprawl, 110–29, 132; urban sustainability, 67, 99, 108, 109, 110, 114, 126, 128, 130–37; water demand and use in, 123–25. *See also* Cities; Environmental Kuznets curve; Metropolitan areas
Urban planning, 109
Urban Water Management Plan Update 2002–2003 (Los Angeles), 49
Uruguay, 68
U.S. *See* United States
Utah, 14
Uzbekistan, 68

Veblen, Thorstein, 58
Vernon, Raymond, 106–07
Virginia, 86, 87

Wackernagel, Mathis, 10
Wang, Hyoung Gun, 97
Washington (D.C.), 114, 120, 121
Washington Metropolitan Area Transit Authority, 121
Water, 107, 123–25. *See also* Environmental issues
Waukegan and Waukegan Harbor (IL), 90–91
Welfare states, 108
Wetlands, 122
WHO. *See* World Health Organization
Wilson, David, 34
Wilson, William Julius, 62
World Bank, 17, 28–29, 32–33, 101
World Development Indicators, 36
World Health Organization (WHO), 1
World Meteorological Organization, 132

Yale University, 67–68

Zagat guides, 28
Zoning, 35, 89, 125–26, 127